The Professional Handbook of

Cider Tasting

Errata

The Professional Handbook of Cider Tasting (Travis Robert Alexander and Brianna L. Ewing Valliere)

The images on the following pages, 8, 16, 54 and 62 taken by photographer Colin Bishop.

The images on the front cover and the following pages, 2, 32, Figure 16 p. 47, Figure 18 p. 56 and Figure 20 p. 59 taken by photographer Jordan Fox and used with permission by the Northwest Cider Association.

Figure 19 p. 57 caption should read Cider and Food Pairing graphic. Used with permission by Northwest Cider Association.

The Cider Institute of North America (CINA) was formed in 2016 in recognition of the need and demand for cider-specific education, training and support for every stage of a career in the cider industry. The Cider & Perry Production Certificate Program offers vocational training in cider and perry production, leading to the achievement of standards defined by the industry and delivered through CINA's training providers including Washington State University, Brock University, and Cornell University. For more information and to sign up for courses visit www.ciderinstitute.com.

The Professional Handbook of
Cider Tasting

Travis R. Alexander and Brianna L. Ewing Valliere

CABI is a trading name of CAB International

CABI	CABI
Nosworthy Way	745 Atlantic Avenue
Wallingford	8th Floor
Oxfordshire OX10 8DE	Boston, MA 02111
UK	USA

Tel: +44 (0)1491 832111
Fax: +44 (0)1491 833508
E-mail: info@cabi.org
Website: www.cabi.org

Tel: +1 (617)682-9015
E-mail: cabi-nao@cabi.org

A catalogue record for this book is available from the British Library, London, UK.

ISBN-13: (paperback) 9781789245493
          (ePDF)   9781789245509
          (ePub)   9781789245516

Commissioning Editor: Rachael Russell
Editorial Assistant: Lauren Davies
Production Editor: James Bishop

Typeset by SPi, Pondicherry, India
Printed and bound in the UK by Bell and Bain, Glasgow

Contents

Preface

The North American cider industry is bigger and more widespread than ever before. Despite the recent growth in cider related research and educational materials, the industry is thirsty as ever for more scientific knowledge to apply in the orchard, the cidery, the tasting room, and beyond. Thanks to the support of the Washington State University Center for Sustainable Agriculture and Natural Resources (CSANR) BIOAg grant program, we are proud to present the first edition of *The Professional Handbook of Cider Tasting*.

The information in this handbook comes from extensive research of existing literature and input from established cider professionals. Our goal has been to produce a manual that provides a comprehensive understanding of cider evaluation principles and methods for both small and large practitioners across North America. However, we recognize that cider is a rapidly expanding and changing industry, and we welcome any feedback for future editions. Questions, comments, and suggestions can be sent to travis.alexander@wsu.edu or bri.ewing@wsu.edu.

This work would not be possible without our industry reviewers. Thank you to Dr. Andrew Lea, Andrew Byers, Dave Takush, Erin James, and Michelle McGrath for your input and commitment to this project. We were continually challenged by your insightful comments, and this manual is much better for it. We would also like to thank the Cider Institute of North America for their partnership and dedication to cider education. Our gratitude also extends to the general cider community including membership organizations such as the United States Association of Cider Makers and the Northwest Cider Association. The collaborative nature of this industry is one of its greatest strengths, and we hope to contribute to the cider industry with our work in return for the immeasurable support that we have received.

Thank you to CABI for publishing this handbook and for recognizing the great potential in the North American cider industry. We appreciate this opportunity to share this information through CABI's globally respected platform. Lastly, on a personal note, Travis Robert Alexander would like to recognize his wife and son, Yael and Jack Alexander, for their unwavering love and patience during the development of this book. Brianna Ewing Valliere would like to acknowledge her husband and best friend, Bryan, for his endless support, compassion, and commitment to sharing laughter and joy together each day.

Acknowledgments

The authors thank the following agencies and organizations for their contribution to this manual: United States Department of Agriculture National Institute of Food and Agriculture, Washington State Department of Agriculture, Washington State University Center for Sustaining Agriculture and Natural Resources, Northwest Agricultural Research Foundation, and Northwest Cider Association.

The authors thank Gary Moulton and Andrew Zimmerman for providing the foundational work from which this manual has resulted, establishing the first cider apple orchard trials and cider sensory evaluations at Washington State University.

(Photo courtesy C. Miles, WSU)

Last but not least, the authors thank review panelists Dr. Andrew Lea (author, cidermaker, and Fellow of the Institute of Food Science and Technology), Andrew Byers (head cidermaker and production manager at Finnriver Farm & Cidery), Erin James (editor-in-chief of CIDERCRAFT magazine), and Dave Takush (head cidermaker and co-owner of 2 Towns Ciderhouse) for their invaluable input in the development of this manual.

(Andrew Lea photo courtesy of John Kruger; Andrew Byers photo courtesy of Finnriver Farm & Cidery; Erin James photo courtesy of Brittany Carvalho Photography; Dave Takush photo courtesy of Atlas Foto)

About the Authors

Travis Robert Alexander is a postdoc in the Department of Horticulture at Washington State University. Travis received his B.S. in Biochemistry from the University of California, San Diego, his M.S. in Horticulture from the University of California, Davis, and his Ph.D. in Horticulture from WSU. His doctorate work focused on advancing cider apple production through evaluations of mechanized harvest and comparisons of regional cider apple juice quality. Travis's postdoctoral work involves evaluating current metrics of cider apple fruit quality to provide growers with a more optimized means of assessing quality up to and at harvest.

Brianna L. Ewing Valliere is a clinical assistant professor in the School of Food Science at Washington State University. Bri received her B.S. in Molecular Environmental Biology from the University of California, Berkeley and her M.S. in Food Science & Technology from Virginia Tech where her master's research focused on the correlation between apple maturity, post-harvest storage conditions, and the quality of cider. Prior to her current position, she gained practical industry experience working at wineries in California and New Zealand. Bri is an instructor partner with the Cider Institute of North America which includes teaching the nationally recognized "Cider & Perry Production – A Foundation" course. She also participates in research to help facilitate the growth of the cider industry.

Introduction

1

Nine years after first landing at Plymouth Rock, European colonists planted apple trees in Massachusetts Bay (Orton, 1995). By the 1670s, orchards in New England were producing up to 500 barrels or a little more than 32,000 gallons of cider annually (Bender, 2009). In 1721, several villages in New England reported cider production of more than 192,000 gallons per year (Williams, 1990). It became common for homesteads to include an apple orchard, and fermentation provided a means to extend the shelf life of apples beyond the growing season (Orton, 1995). Cider was the most popular alcoholic beverage made and consumed in the U.S. until the early 1900s. Mass migration of populations from farms to cities during the Industrial Revolution resulted in the abandonment of orchards and a drop in the supply of fruit. At the same time, mass immigration of beer-drinking Europeans to urban areas provided for the rise of a new beverage culture that was fueled by the availability of inexpensive grain from the Midwest. As a result of Prohibition—a constitutional ban on the production of alcohol for trade from 1920 to 1933—cider's share in the beverage market sharply declined. The production of cider plummeted from a peak of 55 million gallons in 1899 to 13 million gallons in the late 1930s (Watson, 1999). The tradition of cidermaking, which by then was virtually absent on a large commercial scale, would be kept alive merely by farmers and enthusiasts (Orton, 1995).

The rise of the craft beverage movement in the last decade has resulted in cider experiencing a re-birth. From 2012 to 2016, the industry experienced an annualized growth rate of 27.3%, reaching $300.4 million in revenue, and for the next five years it is projected to continue growing at a more sustainable rate of 1.2% (Petrillo, 2016). As of 2016, there were 608 reported cidermakers across 44 states and the District of Columbia, with New York, Michigan, and California fostering more than 60 cidermakers each (Statista, 2017). Washington State, the top dessert apple-producing state in the U.S., experienced a 22-fold increase in the volume of cider produced in the last 10 years, from under 50,000 gallons in 2007 to nearly 1,000,000 gallons in 2016 (Alcohol and Tobacco Tax and Trade Bureau, 2017).

Table 1. Cider apple classification system developed in England at the Long Ashton Research Station (LARS) in Bristol (Barker, 1903).

English classification	Tannin (%)[1]	Acid (%)[2]
Bittersweet	> 0.20	< 0.45
Bittersharp	> 0.20	> 0.45
Sharp	< 0.20	> 0.45
Sweet	< 0.20	< 0.45

[1]Tannin expressed as tannic acid equivalents, in percent.
[2]Titratable acidity expressed as malic acid equivalents, in percent.

Cider apple cultivars in the U.S. have traditionally been categorized according to the English classification system, a system focused specifically on fruit acidity and tannin levels (Table 1). Bittersweets (e.g., "Brown Snout" and "Dabinett") and bittersharps (e.g., "Kingston Black" and "Porter's Perfection") have been the most sought after cultivars for producing high quality, full-bodied ciders (Williams, 1975; Bore and Fleckinger, 1997), but they represent only a minority of what is planted and available to cidermakers across the country. Of the 74 cider apple cultivars maintained in the WSU Mount Vernon NWREC collection, less than half have been classified as a bittersweet or bittersharp per their performance in a maritime climate. Sweets (e.g., "Gravenstein" and "Jonagold") are the least desired cultivars, but currently are the most utilized on a volume basis, as they are most readily available and cost effective (Merwin *et al.*, 2008; Moulton *et al.*, 2010). In Washington State for example, the 2015 average price point for cider apples was $0.375 per lb and for cull dessert apples $0.17 per lb (NWCA *et al.*, 2016).

Quantification of cider apple juice acidity and tannin levels has traditionally been accomplished utilizing titrimetry. Fruit acidity, specifically titratable acidity (TA), can be measured by incrementally adding a strong base such as sodium hydroxide (NaOH) to a sample of diluted juice until a visually confirmed (utilizing a color indicator such as phenolphthalein) or pH monitored (utilizing a pH meter) neutralization point is reached (Fig. 1). TA is measured in units of malic acid equivalents, as malic acid is the predominant organic acid in cider apples. TA in percent malic acid equivalents is quantified using the following equation:

$$\%\mathrm{TA} = \frac{N \times V \times Eq.wt}{v \times 10},$$

where: N = normality of titrant(commonly sodium hydroxide),
V = volume of titrant,
$Eq.wt$ = equivalent weight of acid of interest,
v = volume of sample.

Fruit tannin, specifically titratable tannin (TT), can be measured by incrementally adding the strong oxidizing agent potassium permanganate

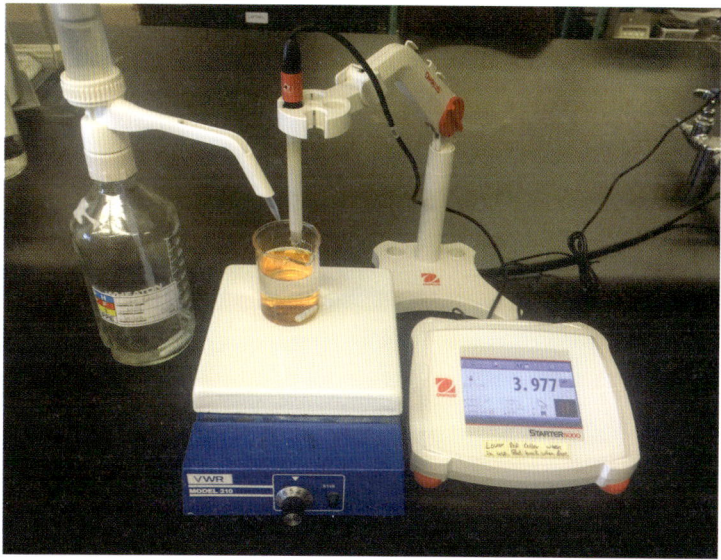

Fig. 1. Titratable acidity measurement setup (left to right): titrant in an electronic burette (0.2 N sodium hydroxide; digitrate 50 mL, Jencons Scientific Ltd, Bridgeville, PA), diluted juice sample (25 mL juice, 100 mL water, and a small magnetic stir bar) on a stir plate (310 Magnetic Stirrer, VWR Scientific, Radnor, PA), and pH meter (Starter 5000 pH Bench, OHAUS, Parsippany, NJ) (photo courtesy T. Alexander, WSU).

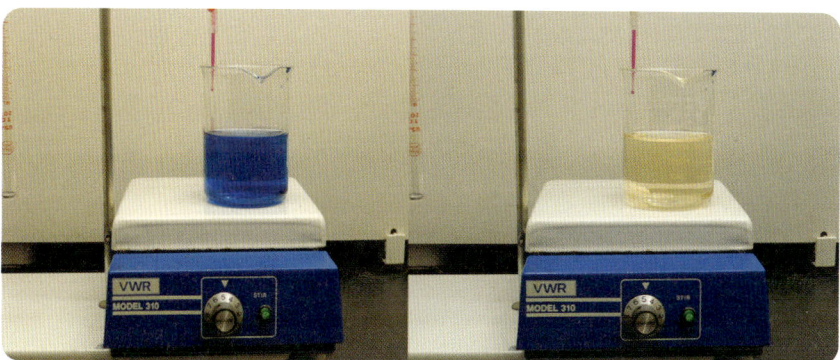

Fig. 2. Titratable tannin assay before and after diluted juice sample (blue; 1 mL juice, 150 mL water, and 5 mL acidified indigo carmine solution) and fully titrated juice sample (yellow; juice, indigo carmine, and titrant), both on a stir plate (310 Magnetic Stirrer, VWR Scientific, Radnor, PA) with a manual glass burette with titrant (pink; 0.005 M potassium permanganate) (photo courtesy T. Alexander, WSU).

(KMnO$_4$) to a sample of diluted juice until a visually confirmed (utilizing the redox indicator indigo carmine) equivalence point (Fig. 2). TT is measured in units of tannic acid equivalents. TT in percent tannic acid equivalents is quantified using the following equation:

$$TT = \frac{\# \, mL \; Titrant \, (sample) - \# \, mL \; Titrant \, (blank)}{10}$$

It is important to note that this TT method provides for a measure of oxidizable compounds in general, delivering an inexact indicator of bitterness and astringency or even of polyphenols in general. Research is underway at universities such as WSU, Cornell, and Virginia Tech to develop an analytical protocol for identifying and quantifying sensory influential tannins in cider apples.

Sensory Evaluation: The Physiological Basis

2

As per the definition established by the Institute of Food Technologists (1975), sensory evaluation is the scientific discipline utilized to evoke, measure, analyze, and interpret responses to those characteristics of foods and materials as they are perceived by the senses of sight, smell, taste, touch, and hearing. The sensory evaluation of cider predominantly involves all the senses except for hearing. It may be argued that the sound of fizz, a popping cork, or clinking glasses could influence the sensory evaluation atmosphere, but hearing is generally not thought to greatly impact the evaluation of cider. The flavor of a cider is a compilation of one's response to its aroma (sense of smell), its taste, and its "mouthfeel" (sense of touch). The appearance of a cider is a product of its color and clarity (sense of sight). It is important to understand how the human senses function to fully appreciate how to conduct an effective evaluation of cider.

Appearance

The sense of sight allows perception of appearance. Through the interaction of the eye, external light, and the brain, humans are able to perceive color and clarity (Fig. 3). Color is processed in terms of hue (tone of color), saturation (intensity of hue from gray to pure), and brightness (relative lightness from black to white). Fig. 4 illustrates these three characteristics of color. Clarity is processed in terms of the degree to which suspended solids are observed (i.e., perceived cloudiness). Though appearance and flavor are two separate quality analyses, consumers may associate appearances with sensory expectations (Engen, 1972). For example, in wine, aroma descriptors are categorized based largely on the color of the wine (Morrot, 2001). Due to these associations with color, there is also the risk of inherent biases. Parr *et al.* (2003) studied the effect of color bias by giving study participants white wines that were colored red in transparent and opaque glassware. Wine experts, though more able to discriminate between white wine aromas and red wine aromas than social drinkers, had more success discriminating

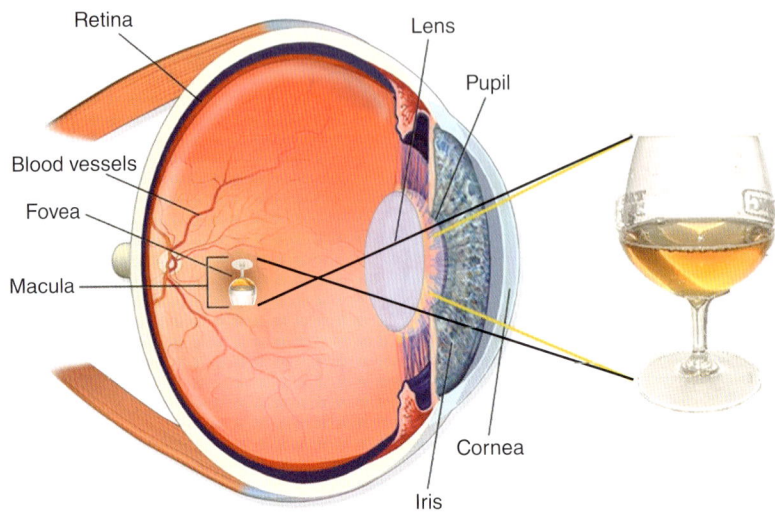

Fig. 3. When light radiates onto a sample of cider, certain wavelengths are reflected to the human eye. These wavelengths travel through the pupil to the lens, which sharpens and inverts the image of the cider sample on the retina. The retina, composed of optic nerves, photoreceptors, and neurons, detects and transmits the inverted image (an impulse) to the brain where it is interpreted and responded to (e.g., brand recognition) (modification of C. Stangor, 2010).

between the wines when the color stimulus was removed by presenting the wine in opaque glassware (Parr *et al.*, 2003). These observations indicate that visual stimuli, such as red coloring, alert the sensory expectations and therefore the experience of the consumer. In cider, the color range is mainly limited to yellow and golden tones, ranging from pale to copper. Though biases regarding color may not be as stark as those in wine, consumers may expect different aromas in a straw-colored cider (e.g., more fresh fruit flavors of a modern-style cider) than in a golden- or amber-colored cider (e.g., more phenolic aromas from a heritage-style cider). Of course, the addition of fruit and spice flavors to modern ciders can convey a wide variety of initial impressions. Taking note of appearance during a sensory evaluation is important for describing the overall quality of a cider.

Aroma

The sense of smell is the most acute, providing for the ability to differentiate thousands of compounds at concentrations as low as parts per trillion. Perception of smell varies greatly among the general population due to genetics, environment, and health. Repeated exposure to aromas provides for

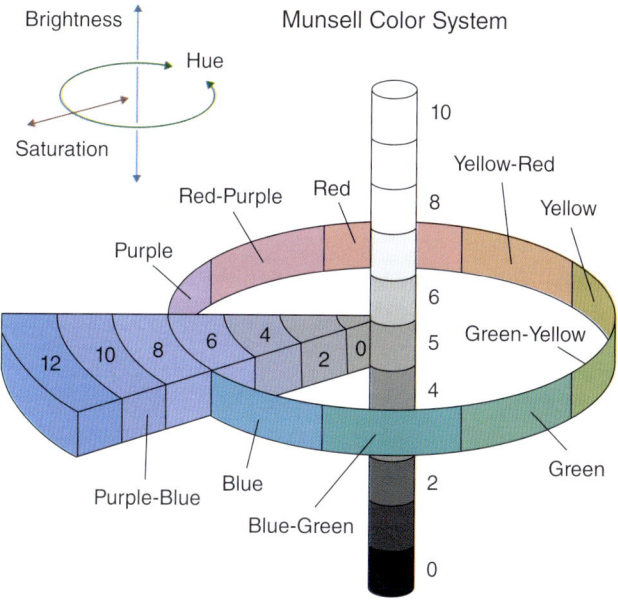

Fig. 4. Diagram of the physiological dimensions of color space with hue varying along the circumference, saturation along the radius, and brightness along the main axis of the proposed imaginary disc (modification of Munsell Color System by J. Rus, 2007; CC-BY-SA 3.0).

an educated response. Training can fine-tune an individual's ability to differentiate compounds, and geographical origin can cause varied responses to stimuli due to exposure to different aromas throughout life. Aromas perceived in cider may originate from multiple sources and accordingly are divided into three categories: primary aromas, secondary aromas, and tertiary aromas. Primary aromas originate from the fruit itself, secondary aromas originate from the cidermaking process, including yeast or bacterial metabolic processes (e.g., fruit esters from yeast metabolism; buttery diacetyl notes from malolactic fermentation), and tertiary aromas originate from maturation and aging (e.g., toasty notes from oak barrels) (Bakker and Clarke, 2011). Therefore, each production decision, ranging from fruit selection to maturation vessel selection, can have a significant effect on the overall bouquet of a cider.

Taste

The sense of taste has been simplified by five discrete descriptors: bitter, salty, sour, sweet, and umami (Smith and Margolskee, 2001; Jackson, 2009). Bitter is often used to describe quinine (key ingredient of tonic water), while sour may be used to describe lemons or unripe citrus. Colloquially, bitterness and sourness are often confused, but each term should only be used to describe its specific taste. Bitterness is the product of procyanidin polar moieties binding with taste receptors in the papillae membranes of the tongue, and perception generally decreases with time (a process commonly termed "softening"; Koyama and Kurihara, 1972). Chapter 5 will introduce and elaborate procyanidins in terms of their origins and interactions with other molecules as this impacts the final product. "Sweet" can be used to describe sugary confections, "salty" can used to describe table salt, and umami used to describe the savory flavor of meat. The taste of cider also involves the descriptor of "metallic," a perception, of which the physiological basis is still not completely understood, that can be replicated by oral contact with a metal spoon. Metal ions such as copper and iron have been found to stimulate metallic taste at concentrations not commonly found in cider (> 20 and 2 mg/L, respectively), and detection is amplified by the presence of tannins (Moncrieff, 1964). While researchers currently recognize five basic tastes, humans experience a range of complex flavors as a result of the interaction of smell and taste. Volatile compounds can travel not only through the orthonasal pathway (i.e., nasal cavity; Fig. 5), but also through the retronasal pathway at the back of the mouth providing for simultaneous stimulation of taste and smell (Fig. 6). This latter interaction can be hindered by the unintentional (e.g., illness) or intentional (e.g., squeezing the nose) blockage of the nasal system which results in a perceived blandness, lack of taste.

Mouthfeel

"Mouthfeel" is the sensory and functional manifestation of structural, mechanical, and surface properties of food detected through the action of touching and influenced by sight and hearing (Fig. 7). Mouthfeel is the perceived texture in the mouth and is influenced by flow characteristics, resistance to mastication, and temperature. Body, astringency, and fizziness are all tactile terms that are utilized to describe the weight, dryness, and prickliness of a cider sample in the mouth. Pectic polysaccharides that are present in both grapes and apples have been shown to contribute to the fullness sensation in wines (Vidal *et al.*, 2004). Astringency in cider is the product of hydrogen bonding between procyanidins and proteins of the tongue, and perception of astringency increases with the capacity for hydrogen bonding (i.e., with larger-sized procyanidins; Joslyn and Goldstein, 1964). Just as taste and smell interact with one another to produce more complex interactions, mouthfeel properties can also influence other senses. For example, temperature at which a cider is served is important because at cooler temperatures (41 to 59 °F) the perception of acidity and sweetness along with aromas are reduced, while the perception of bitterness and astringency are amplified.

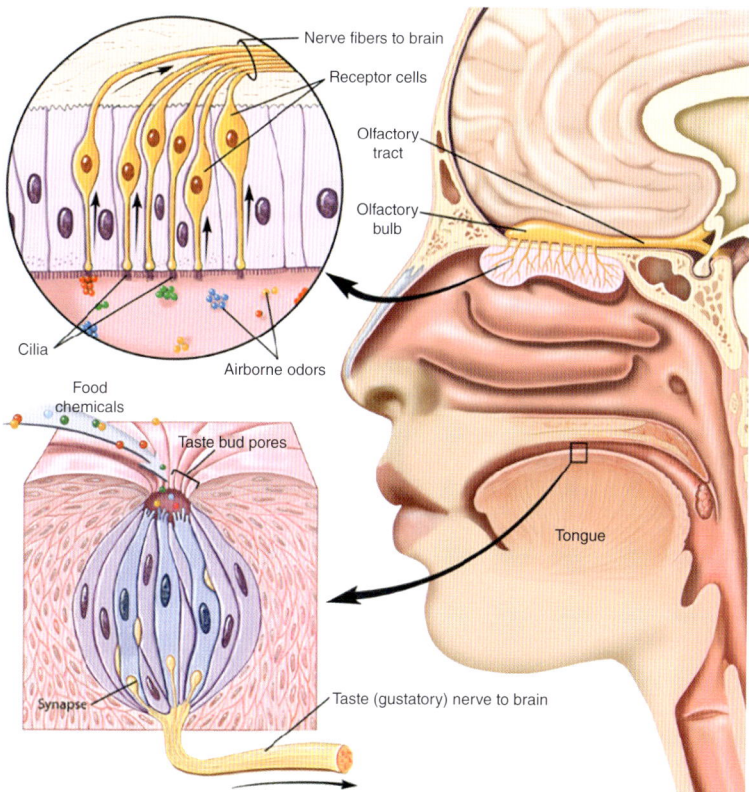

Fig. 5. To be tasted, a compound must be solubilized so that it can either bind with taste bud receptors (bitter, sweet, and umami compounds) or interact with ion channels within the taste buds (salty and sour compounds). Binding with receptors or electrical interaction with ion channels triggers signal transmission to the brain resulting in the manifestation of a physiological response (e.g., recognition of sweet taste in response to the presence of saccharides) (Stein and Stoodley, 2006).

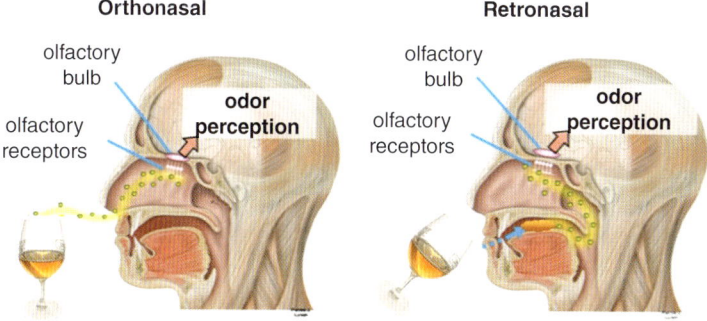

Fig. 6. Orthonasal and retronasal routes to odor perception, the former involving the inhalation of stimuli through the nasal cavity and the latter inhalation through the back of the mouth (Lynch, 2006).

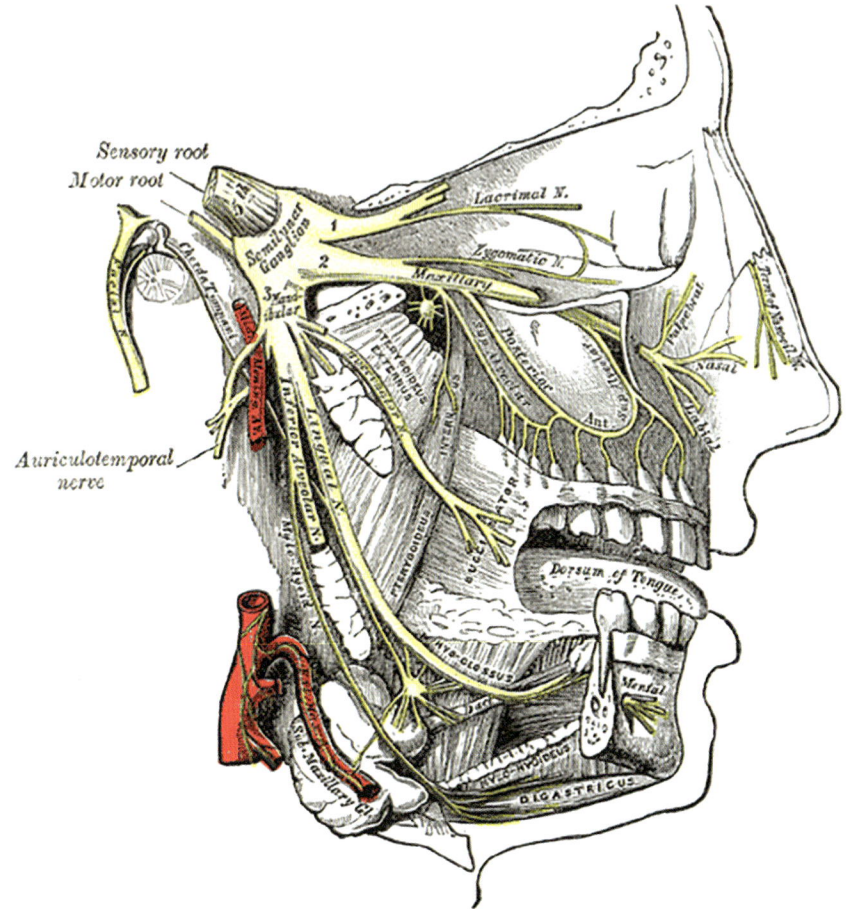

Fig. 7. Three types of receptors underlie perception of texture: mechano-receptors, thermoreceptors, and nociceptors. Mechanoreceptors provide for a response to pressure, thermoreceptors provide for a response to temperature, and nociceptors provide for a response to chemical irritation. The trigeminal nerve is responsible for transmitting information from these receptors in the eyes, nose, and mouth, to the brain. Ultimately, signaling to the brain results in physiological responses (e.g., vasodilation and bronchodilation) (Gray, 1918)

Sensory Evaluation: Effective Profiling

3

Sensory evaluation of cider is conducted to obtain either objective or subjective data. The collection of objective data involves technical analysis of the sensory attributes of the cider for the purpose of developing and/or monitoring quality, as commonly performed by cidermakers. Quantitative assessments can involve a panel of trained individuals or calibrated instruments. The collection of subjective data involves the non-technical evaluation of a cider's sensory attributes for the purpose of developing and/or informing on preference, as commonly performed by cider consumers. Qualitative assessments are heavily dependent on experience that with augmentation, provides the knowledge to recognize and meaningfully describe attributes.

Setup

Whether one plans to perform a quantitative or qualitative assessment, preparation is key to effective profiling. To the extent possible, all sources of background stimulation should be managed. Fragrances or perfumes should be avoided. Music and communication should be kept to a minimum. Lighting should be adequate, and decorum not overwhelming. An odor-free, quiet, bright, and neutral-colored tasting room is the optimal setup. Food should be consumed in moderation pre-evaluation, as alcohol is absorbed by the body quicker on an empty stomach, and the senses of smell and taste are less acute on a full stomach. Cider should be consistently presented in proper glassware in a size and style that is appropriate for the type of cider being poured. In general, glass capacity should be around 8 to 12 oz with up to one-third of the glass filled per sample, and glasses should have a shape that concentrates aromas and minimizes splashing (e.g., tulip-shaped). While effective for marketing, large and bold logos on glasses should be avoided as they can be distracting. Cider can also be poured into crystal stemware, which has a more slender construction than glass but at a higher cost. For many formal beverage evaluations, there are standard tasting glasses that follow specific dimensions in order to create uniformity in sensory evaluations. In wine tasting, for example, the standard wine glass is the International Organization for Standardization (ISO) specification 3591:1977 (iso.org). In terms of use, vessels should be sufficiently cleaned between samples to avoid residues from previous samples. To further avoid sensory fatigue, tasters should be provided with water for rinsing, waste containers for disposal of sample, and bread or crackers for palate refreshment. Finally, if multiple types of ciders are to be evaluated in one sitting, lighter-bodied, younger aged, and less tannic ciders should be evaluated first. In addition, the samples should be presented from dry to sweet.

Presentation

The number of ciders evaluated in a sitting should be adjusted in response to taster fatigue. During sensory assessment of a sample, communication regarding perception should be avoided until after a tasting has been completed. If conducting a blind tasting, the identity of cider should be kept unknown until after a tasting has been completed. A bottle can be covered with a paper bag and labeled with a random three-digit code (Fig. 8). Avoid single-digit and consecutive number coding to prevent bias. Bias should be minimized by excluding information such as reputation, price, origin, or availability.

In utilizing the four senses previously explained, ciders are generally sequentially assessed in terms of appearance, smell, taste, mouthfeel, and finish (Fig. 9). When assessing appearance, a glass of sample should be held

Fig. 8. Cider samples labeled with three digit random codes to minimize evaluator bias (photo courtesy T. Alexander, WSU).

1	Observe
2	Smell
3	Taste
4	Feel
5	Swallow or Spit

Fig. 9. Five steps to evaluating cider as modified by the authors (Henderson and Rex, 2012).

up to a light source for optimal assessment of clarity, and it should be held against a white surface for optimal assessment of color. When assessing aroma, the assessor should first smell the sample without swirling and take note of the aromas experienced. Then, the assessor should swirl the glass and gently sniff for additional aromas that may have been released as a result of swirling. When assessing taste and mouthfeel, the assessor should take a small sip of cider and hold the sample in the mouth for a couple of seconds to allow for perception of acidity, sweetness, bitterness, viscosity, alcohol content, astringency, etc. While the cider remains in the mouth, air can be drawn into the mouth in a slurping-like manner. This introduction of air agitates the sample and allows for a greater release of aromatic compounds. Additional sips can be taken to confirm first impressions and to allow for assessment of finish and overall balance, such as the ratio of acidity to sweetness. Observations should be recorded at each step of the sensory assessment and shared respectfully and objectively for the development of skill. Tasters can be supplied with a tasting sheet (Fig. 10) to record notes and/or provide feedback. Individual sensitivity to different aromas

Cider & Perry Organoleptic Profile Form								
Product:								
Assessor:						**Date:**		
Appearance	**Description**							
Clarity								
Color								
Carbonation								
Brightness								
Aroma								
Aroma	**Intensity**							
Description	1	2	3	4	5	6	7	
Taste								
Taste	**Intensity**							
Description	0	1	2	3	4	5	6	7
Sweet								
Sour								
Bitter								
Salty								
Mouthfeel								
Mouthfeel	**Intensity**							
Description	0	1	2	3	4	5	6	7
Body								
Astringency								
Length/Finish								
Other								

Fig. 10. Tasting sheet utilized by the WSU Cider Education Program (B. Valliere, WSU. Adapted from P. Mitchell, 2016. *Cider & Perry Production – A Foundation. The Official textbook for the Foundation Certificate in Cider and Perry Production.*).

and tastes (i.e., thresholds for detection) will be established with practice, but standardization via the use of cider lexicon should be encouraged from the beginning (Table 2).

Training

Selection and training of assessors, commonly referred to as "tasters," is the first step in conducting a quantitative assessment. Each individual of a tasting panel should be self-motivated to participate, have above average familiarity with the product, be of generally good health, and contribute to an overall fair representation of the target consumer group. Panelists should be committed to the entirety of the session. Motivation to complete the assessment may be achieved by establishing clear objectives, providing a comfortable environment, and presenting financial incentives (e.g., food vouchers), if appropriate. Extensive experience with cider is advantageous as it allows for a more advanced training and provides for panelists with greater confidence in their ability to accomplish the task at hand. Health issues or habits that negatively impact the senses, such as the common cold, color blindness, or smoking, should be screened against in selecting panelists. Putting together a panel that represents the target consumer group is essential to ensuring that the assessment has value. Those attributes preferred by cider enthusiasts typically differ from the average cider consumer, but consumers as a whole are far from a homogenous group.

Table 2. Cider and perry descriptors (P. Mitchell, Mitchell F&D Limited, 2016).

Main Descriptors (Cider & Perry)		

• Appearance
  A. Clear; Cloudy; Bright
  B. Straw; Amber;
    Golden; Copper
• Aromatic
  A. Cidery
  B. Winey; Pear Drops;
    Estery; Floral
  C. Spirituous; Piquant
• Herbaceous
  A. Grassy; Elderflower;
    Vegetative
  B. Hay-straw
  C. Nutty
• Yeasty
  A. Yeasty
	• Fruity
	  A. Bittersweet Apple;
	    Culinary; Pear
	  B. Tropical fruits
	  C. Summer fruit
	  D. Berry fruits
	  E. Citrus fruits
	  F. Dried Fruit; Cooked
	    Fruit
	• Spicy & woody
	  A. Spicy
	  B. Woody
	  C. Phenolic
	• Sweet Associations
	  A. Caramel;
	    Butterscotch; Vanilla
	  B. Honey; Syrup
	  C. Confectionary
		• Primary tastes
		  A. Sweet
		  B. Acid
		  C. Salty
		  D. Bitter
		• Mouthfeel
		  A. Body
		  B. Astringency;
		    Creamy;
		    Warming;
		    Carbonation;
		    Metallic; Powdery
		• Overall flavour
		  A. Fullness
		  B. High/Low
		    Intensity
		  C. Balanced
		  D. After-taste

While there does not exist a single test for identifying superior tasters, there are various tests for measuring indicators of tasting capacity. These tests can demonstrate the strengths, weaknesses, and the overall individual aptitude for sensory analysis. Many tests are available that require relatively low financial and intellectual inputs compared to those implemented by researchers, and smartphone apps for running sensory panels and analyzing data are now more available, providing easily accessible tools and resources

Taste recognition and acuity

Prepare two sets of sensory standards (Table 3). Prepare the solutions an hour before tasting, stir sufficiently to complete dissolution, and allow potential panelists to taste the two bases (water and cider) for familiarity. First evaluating set 1 and then set 2, present 30 mL of each solution in random order and have all taste and texture attributes detected, along with their intensities, recorded on a tasting sheet (Fig. 11). The ability to recognize taste and texture stimuli are examined in the context of water and cider as perception can differ significantly with solvent, as shown with wine in Fig. 12.

Table 3. Sensory standards for testing recognition of taste and texture sensations.

Standard	Set 1 (in 750 mL water)	Set 2 (in 750 mL cider)[1]
Astringent	1 g tannic acid	1 g tannic acid
Bitter	10 mg quinine sulfate	10 mg quinine sulfate
Hot	50 mL ethanol	50 mL ethanol
Sour	2 g malic acid	2 g malic acid
Sweet	15 g sucrose	15 g sucrose

[1]Utilize a commercial product widely consumed in your market.

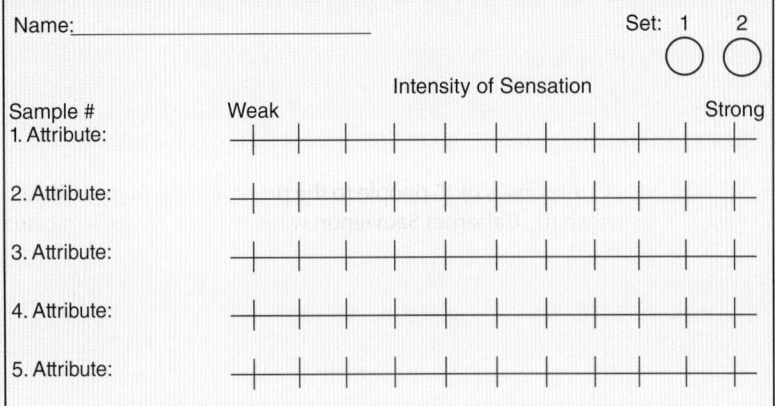

Fig. 11. Tasting sheet for acuity test.

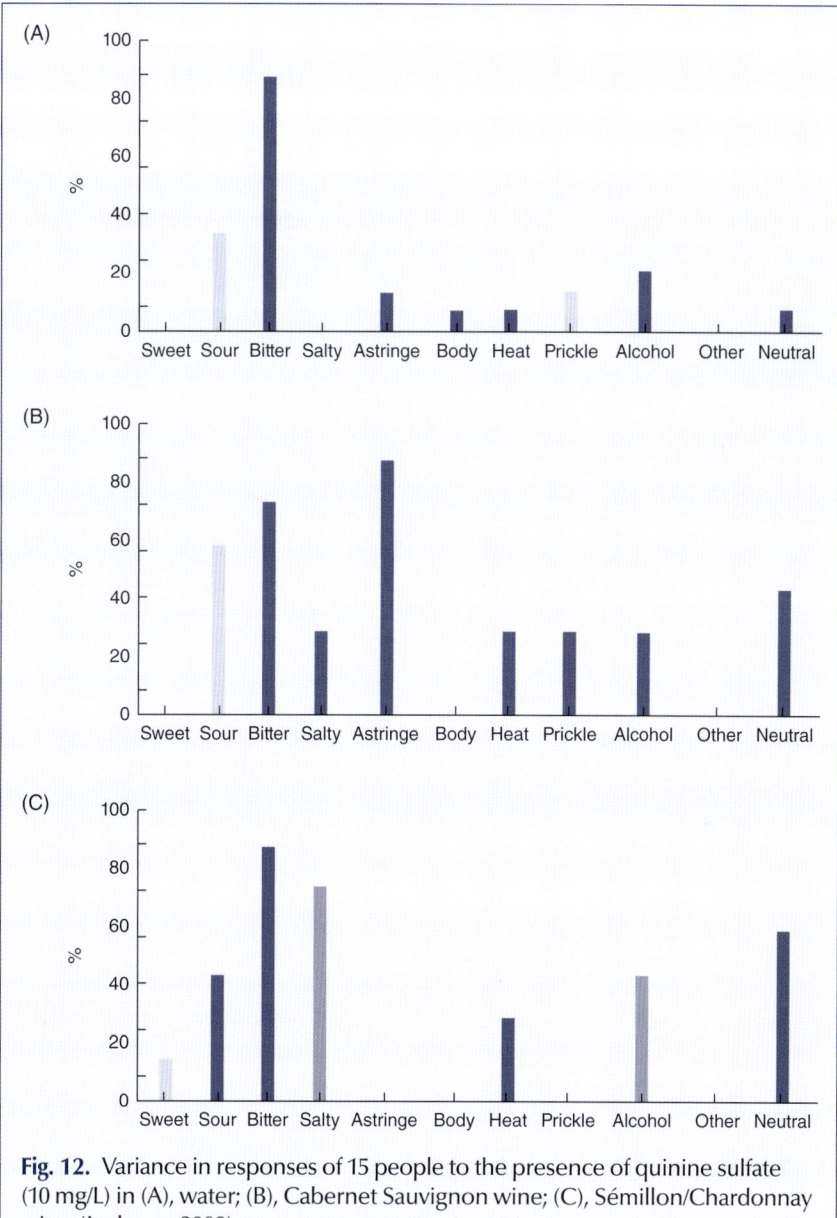

Fig. 12. Variance in responses of 15 people to the presence of quinine sulfate (10 mg/L) in (A), water; (B), Cabernet Sauvignon wine; (C), Sémillon/Chardonnay wine (Jackson, 2009).

to cider producers. Lawless and Heymann's (2010) *Sensory Evaluation of Food Principles and Practices* (second edition) provides more comprehensive insight into the statistical analyses that can be utilized to mathematically confirm what is self-evident from observation of data. In terms of the optimal number of tasters, the more panelists that can be trained the greater the probability that collected data will reflect the consumer group the panelists have been chosen to represent. At the same time, more panelists equate to a greater cost because of the time and supplies it takes to put together a panel. In practice, a larger operation should have a trained pool of 15–20 with 10–12 reliable tasters and a smaller operation should have a pool of 10–12 tasters with 6–8 dependable members.

Relative taste sensitivity

To assess relative perception of an attribute, present a series of cider samples with varying concentrations of an attribute standard: astringency (0, 5, 10, 20, 40 g/L tannic acid); bitterness (2.5, 5, 10, 20 mg/L quinine sulfate); sourness (0, 5, 10, 20, 40 g/L malic acid); sweetness (0, 2.5, 5, 10, 20 g/L sucrose). To maintain sample independence, provide potential panelists with unsalted crackers and water for palate cleansing and request a 30-second pause between sampling for palate reset. Also, have the participants familiarize themselves with the base cider. Present each person with the randomly numbered samples for ranking, in order from weakest to strongest sensitivity. The test should be repeated with different types of cider (Table 4).

Table 4. Sourness sensitivity test utilizing three different ciders.

Cider 1	Cider 2	Cider 3
#3[1] (A)[2]	#5 (C)	#5 (E)
#2 (B)	#4 (D)	#1 (B)
#5 (D)	#1 (A)	#3 (D)
#1 (E)	#2 (B)	#4 (C)
#4 (C)	#3 (E)	#2 (A)

[1]# 1–5 label identifying the sample.
[2]A = control, 0 g/L (malic acid); B = 5 g/L; C = 10 g/L; D = 20 g/L; E = 40 g/L.

Aroma recognition: fragrances and faults

Prepare aroma standards (Table 5 and 6 or Table 7) and then allow potential panelists to study the standards along with a control (unmodified base cider), taking notes on aromas that differentiate the standards and the control. Ensure that standards are constantly covered when not being assessed (e.g., 60 mm plastic petri dish film) to prevent the loss of aromas and also to allow for vigorous swirling before assessment. For the formal assessment, present the

continued

standards along with controls in dark-colored glasses, controls helping to minimize identification by elimination. Utilizing a response sheet that lists descriptor terms and sample numbers, instruct participants to mark the sample number or control opposite the descriptor term that was perceived. The formal assessment should be repeated three times to allow for adequate examination of a panelist's ability to not only identify, but also learn, new aromas.

Table 5. Preparation of aroma fragrance standards for aroma recognition testing by research labs or large commercial operations (Jackson, 2009).

Sample	Amount per 300 ml of base wine[a]
Temperate tree fruit	
Apple	15 mg Hexyl acetate
Cherry	3 ml Cherry brandy essence (Noirot)
Peach	100 ml Juice from canned peaches
Apricot	2 Drops of undecanoic acid γ-lactone plus 100 ml juice from canned apricots
Tropical tree fruit	
Litchi	100 ml Litchi fruit dirink (Leo's)
Banana	10 mg Isoamyl acetate
Guava	100 ml Guava fruit drink (Leo's)
Lemon	0.2 ml Lemon extract (Empress)
Floral	
Rose	6 mg Citronellol
Violet	1.5 mg β-Ionone
Orange blossom	20 mg Methyl anthranilate
Iris	0.2 mg Irone
lily	7 mg Hydnoxycitronellal
Vegetal	
Beet	25 ml Canned beet juice
Bell pepper	5 ml 10% Ethanolic extract from dried bell pepper (2 g)
Green bean	100 ml Canned green bean juice
Herbaceous	3 mg 1-Hexen-3-ol
Spice	
Anise/licorice	1.5 mg Anise oil
Peppermint	1 ml Peppermint extract (Empress)
Black pepper	2 g Whole black peppercorns
Cinnamon	15 mg *trans*-Cinnamaldehyde
Nuts[b]	
Almond	5 Drops bitter almond oil
Hazelnut	3 ml Hazelnut Essence (Noirot)
Coconut	1.0 ml Coconut Essence (Club House)

continued

Table 5. Continued.

Sample	Amount per 300 ml of base wine[a]
Woody	
Oak	3 g Oak chips (aged ≥ 1 month)
Vanilla	24 mg Vanillin
Pine	7.5 mg Pine needle oil (1 drop)
Eucalyptus	9 mg Eucalyptus oil
Pyrogenous	
Incense	half a Stick of Chinese incense
Smoke	0.5 ml Hickory liquid smoke (Colgin)
Mushroom	
Agaricus	Juice from 200 g microwaved mushrooms
Truffle	30 ml Soy sauce
Miscellaneous	
Chocolate	3 ml Chocolate liqueur
Butterscotch	1 ml Butterscotch flavor (Wagner)

[a]With whole fruit, the fruit is ground in a blender with 95% alcohol. The solution is left for about a day in the absence of air, filtered through several layers of cheesecloth, and added to the base wine. Several days later, the sample may need to be decanted to remove excess precipitates.
Important: All participants should be informed of the constituents of the samples. For example, people allergic to nuts may have adverse reactions even to their smell.

Table 6. Preparation of aroma fault standards for aroma recognition testing by research labs or large commercial operations (Jackson, 2009).

Sample	Amount (per 300 ml neutral-flavored base wine)[abcd]
Corked	
2,4,6-TCA	3 µg 2,4,6-trichloroanisole
Guaiacol	3 mg Guaiacol
Actinomycete	2 mg geosmin (an ethanolic extract from a *Streptomyces griseus* culture[e])
Penicillium	2 mg 3-octanol (or an ethanolic extract from a *Hemigera* (*Penicillium*) culture[f])
Chemical	
Fusel	120 mg Isoamyl and 300 mg isobutyl alcohol
Geranium-like	40 mg 2,4-Hexadienol
Buttery	12 mg Diacetyl[g]
Plastic	1.5 mg Styrene
Sulfur	
Sulfur dioxide	200 mg Potassium metabisulfite
Goût de lumière	4 mg Dimethyl sulfide[g] and 0.4 mg ethanethiol

continued

Table 6. Continued.

Sample	Amount (per 300 ml neutral-flavored base wine)[abcd]
Mercaptan	4 mg Ethanethiol
Hydrogen sulfide	2 ml Solution with 1.5 mg $Na_3S.9H_2O$
Miscellaneous	
Oxidized	120 mg Acetaldehyde
Baked	1.2 g Fructose added and baked 4 weeks at 55°C
Vinegary	3.5 g Acetic acid
Ethyl acetate	100 mg Ethyl acetate
Mousy	alcoholic extract culture of *Brettanomyces* (or 2 mg 2-acetyltetrahydropyridines)

[a]To limit oxidation, about 20 mg potassium metabisulfite may be added per 300 ml base wine.

[b]Because only 30 ml samples are required at any one time, it may be convenient to disperse the original sample into 30 ml screw-cap test tubes for storage. Parafilm can be stretched over the cap to further prevent oxygen penetration. Samples stored in a refrigerator usually remain good for several months.

[c]Other off-odor sample preparations are noted in Meilgaard *et al.* (1982).

[d]**Important:** participants should be informed of the chemical to be smelt in the test. For example, some asthmatics are highly sensitive to sulfur dioxide. If so, such individuals should not serve as wine tasters.

[e]*Streptomyces griseus* is grown on nutrient agar in 100 cm diameter petri dishes for 1 week or more. The colonies are scraped off and added to the base wine. Filtering after a few days should provide a clear sample.

[f]*Penicillium* sp. isolated from wine corks is inoculated on small chunks (1–5 mm) of cork soaked in wine. The inoculated cork is placed in a petri dish and sealed with Parafilm to prevent the cork from drying out. After 1 month, obvious growth of the fungus should be noticeable. Chunks of the overgrown cork are added to the base wine. Within a few days, the sample can be filtered to remove the cork. The final sample should be clear.

[g]Because of the likelihood of serious modification of the odor quality of these chemicals by contaminants, Meilgaard *et al.* (1982) recommend that they be purified prior to use: for diacetyl, use fractional distillation and absorption (in silica gel, aluminum oxide, and activated carbon); for dimethyl sulfide, use absorption.

Once panel selection is completed, panelists can be trained to recognize sensory attributes of cider. Tasters should be supplied with a tasting sheet to record and submit evaluations. Chemical standards (e.g., Cider Sensory Kit – Series One, FlavorActiV, Oxfordshire, UK) or freshly prepared standards (Fig. 13; Table 7) can be utilized for the development and fine-tuning of perceptions. It is imperative that communication is accurate and clear throughout the training process. The strict use of cider terminology, which will be developed over the course of training, is important in minimizing

Fig. 13. Preparation of fresh cider standards (photo courtesy of T. Alexander, WSU).

variation in sensory results due to panelist response differences rather than sample differences. The ultimate goal of training is to reach a point where panelists agree on a common vocabulary and list of attributes that can provide for a panel that is comfortable in providing consistent and accurate evaluation.

Formal Assessment

Once there is confidence in the panelists' ability to evaluate samples consistently and accurately, the next step is formal evaluation. As discussed previously, the optimal evaluation setting should have adequate lighting, be painted in neutral colors with matt-white tabletops, be well ventilated, free of distracting noises, and, if economically feasible, designed such that panelists are physically isolated to avoid interaction. Each panelist should be provided with a pitcher of water and unsalted crackers or white bread for palate cleansing. There is no ideal number of samples that should be presented to a panelist within a session, but with a greater number of samples a panelist is more prone to sensory fatigue. The evaluation of replicates is useful in terms of accounting for panelist variation and having substitutes on hand, but often they are unavailable for economic reasons. Sufficient screening and

Table 7. Aroma, flavor, mouthfeel, and taste attributes (principal and secondary terms), reference standards, base solutions of standards, and referenced intensities that can be utilized in non-research laboratory evaluation of ciders (Alexander *et al.*, 2017).

Principal term	Secondary term	Aroma standard[1]	Base solution[2]	Intensity (cm)[3]
Burnt	Caramel	35 g caramelized sugar	CHC	10
Chemical	Ethanol	15 mL 950 g kg⁻¹ ethanol	CHC	9
Earthy	Earthy	20 g potting soil	CHC	11
Floral	Rose	3 drops 2-phenylethanol	CHC	10
Fruity	Apple	chopped ½ "Granny Smith"	CHC	10
Fruity	Citrus	½ grapefruit (*Citrus ×paradisi*)	CHC	9.5
Herbaceous	Grassy	2 drops hexanal	CHC	9
Microbiological	Yeasty	1 tbsp active dry yeast	CHC	10.5
Spicy	Clove	10 clove (*Syzygium aromaticum*)	CHC	11
Woody	Oaky	12 oak (*Quercus alba*) chips soaked overnight	CHC	9.5

Principal term	Secondary term	Flavor standard	Base solution	Intensity (cm)
Burnt	Caramel	35 g caramelized sugar	CHC	9
Earthy	Earthy	20 g potting soil	CHC	11
Floral	Rose	3 drops 2-phenylethanol	CHC	10
Fruity	Apple	chopped ½ "Granny Smith"	CHC	9
Fruity	Citrus	½ grapefruit with rind	CHC	11
Herbaceous	Grassy	2 drops hexanal	CHC	9
Microbiological	Yeasty	1 tbsp active dry yeast	CHC	10.5
Spicy	Clove	10 whole clove	CHC	10.5
Woody	Oaky	12 oak chips soaked overnight	CHC	9.5

Principal term	Secondary term	Mouthfeel standard[4]	Base solution	Intensity (cm)
Drying	Astringent	0.10 g alum sulfate	CHC	9
Prickly	Carbonated	20 mL mineral water	None	10

continued

Table 7. Continued.

			None	Low/High
Mouth coating	Creamy	Refer to definition	None	Low/High
Heat	Ethanol	35 mL ethanol	H_2O	10
Metallic	Metallic	6 ground iron tablets	CHC	9.5
Principal term	**Secondary term**	**Taste standard[5]**	**Base solution**	**Intensity (cm)**
Bitter	Bitter	0.025 g quinine sulfate	AJ	11
Acidity	Sour	2 g malic acid	AJ	11
Sweet	Sweet	2.5 g sucrose	AJ	9

Finish

Term	Definition
Duration	Length taste lasts in the mouth. Low: 0–20 seconds Medium: 20–40 seconds High: 40–60+ seconds
Detection of: Apple, astringent, metallic and sweet	Detection (0 = no, 1 = yes) of descriptors, as defined previously, after spitting out of sample.

[1]Active dry yeast (Red Star Yeast Co., Milwaukee, WI), whole cloves (McCormick & Co., Sparks, MD), fresh fruit (Safeway, Pullman, WA), sucrose (California and Hawaiian Sugar Co., Crockett, CA), ethanol, hexanal, and 2-phenlyethanol (Sigma-Aldrich, St. Louis, MO), potting soil (Scotts Miracle-Gro, Marysville, OH), and oak chips (Gusmer Enterprises, Mountainside, NJ).

[2]CHC = base solution of 250mL commercial cider (Crisp Apple; Boston Beer Co., Boston, MA), H_2O = base solution of 250 mL filtered water (EcoLab, St. Paul, MN), and AJ = base solution of 250 mL apple juice (Tree Top, Prosser, WA).

[3]1 cm = 0.3937 inch, 1 g = 0.0353 oz, 1 mL = 0.0338 fl oz, 1 $g \cdot kg^{-1}$ = 0.0160 oz/lb, 1 drop = 0.05 mL, 1 tablespoon (tbsp) = 14.7868 g

[4]Alum sulfate (McCormick & Co.), mineral water (Perrier, Vergèze, France), and iron tablets (Nature Made, Northridge, CA).

[5]Quinine sulfate (Sigma-Aldrich) and malic acid (Brewcraft, Vancouver, WA).

panelist training should help minimize the need for replication. There does not exist a general consensus regarding the optimum temperature for evaluating cider, merely scientifically-based guidelines. On the one hand, serving cider chilled is most representative of what the consumer will be experiencing. However, as you warm the cider, more aroma compounds will begin to be released due to the equilibrium of the volatile and non-volatile phases. In choosing a temperature at which samples will be presented, one will have to make sensory sacrifices. Cool temperatures, such as the general suggested white wine serving temperature of 50 °C, for example, enhance the prickling sensation, impart a refreshing sensation, increase the metallic sensation, mellow intense bouquets, decrease the burning sensation of high alcohol content, and reduce sensitivity to sugars.

Decanting samples can be valuable if there is a significant presence of sediment and allows for the early detection of aromatic faults. Decanting does expose the sample to potential oxidation which can impact the sensory profile if sampling does not occur immediately, and decanting of sparkling or carbonated ciders may significantly impact the carbonation level at tasting. Depending on the objectives, sample volume should range from 30 to 70 mL, and samples should be evaluated at a time when panelists are most alert. In conclusion, human panelists are prone to bias that can be mitigated with proper preparation, practice, and presentation.

List of sensory evaluation errors

Adaptation. Change in sensitivity (generally reduction) due to repeated exposure to a stimulus or interaction of stimuli. Limiting sample number and size, refreshing the palate in between samples, and pausing for a duration of time in between samples can inhibit this effect.

Error of habituation. The tendency to rate a series of samples presented in increasing or decreasing stimulation as the same. Randomization of samples presented can prevent the occurrence of this error.

Error of expectation. The triggering of preconceived notions in response to information (verbal or written) presented with a sample. Planning and practice by the panel administrator can reduce the occurrence of this error.

Halo effect. The tendency in which the evaluation of one attribute influences the impression of another when assessing multiple attributes at one time. Separation of evaluations (e.g., preference and descriptive profiling) can inhibit this effect.

Logical error. The association of two or more characteristics of a sample in the mind of the assessor and modification of his or her verdict despite actual perception. Uniformity of samples, masking of differences, and training can prevent this error.

Types and Styles of Cider

4

It is human nature to create labels and categories for objects and ideas. Sometimes, this categorization is rather vague, while at other times categorization can be highly specific. With cider, there currently exist three levels of differentiation: type, style, and class. While there is the potential for further differentiation, only the broadest category will be of value to those with little experience (e.g., first-time consumers). The following are style guidelines modified from those developed by the Great Lakes International Cider and Perry Competition (GLINTCAP; version 2018; Grand Rapids, MI), the world's largest cider judging event as of 2018. The United States Association of Cider Makers has also developed a style guide, but it is less technical and more focused on the consumer than a producer or researcher.

Standard

Modern ciders

Made primarily from culinary (e.g., "Bramley's Seedling") and dessert apples (e.g., "McIntosh"). Compared to other standard styles, these ciders are generally lower in tannin and higher in acidity.

Aroma/flavor. Sweet or low-alcohol ciders may have apple aroma and flavor. Dry ciders will be more wine-like with some esters. Acidity is refreshing, but must not be harsh.

Appearance. Brilliant; pale to yellow in color.

Mouthfeel. Medium body.

Overall Impression. Refreshing drink that is not bland or watery. Sweet ciders must not be cloying. Dry ciders must not be too harsh.

Traditional ciders

Encompass those produced in the West Country of England (notably Somerset and Herefordshire), northern France (notably Normandy and Brittany), the Asturias and Basque Country, and other regions in which cider-specific apple varieties and production techniques are used to achieve a profile similar to English, French, and Spanish ciders. In Asturias, these ciders are known as *sidra natural*. In the Basque Country, these ciders are known as *sagardo naturala*. Made primarily from multi-use (e.g., "Braeburn"), cider-specific apples (e.g., "Brown Snout," a bittersweet, and "Kingston Black," a bittersharp), and/or wild or crab apples (e.g., "Hewe's Crab") (sometimes used for acidity/ tannin balance). These ciders will generally be higher in tannin than modern ciders. Most ciders in the traditional English style will be in the dry class (sweetness level is dry or medium-dry), while most ciders in the traditional French style will be in the sweet class (sweetness level is medium to sweet).

Aroma/flavor. English-style may have no overt fresh apple character but instead various flavors and esters that suggest fermented apple. English-style ciders commonly go through malolactic fermentation (MLF), which results in lowered perceived acidity and possibly production of aromatics such as diacetyl. Spicy, smoky, and phenolic aromas result from the English cider apples, and farmyard/horse characters may be an indication of *Brettanomyces*. These flavor notes are positive but not required. If present, they must not dominate; in particular, the phenolic and farmyard notes should not be heavy. French-style ciders will have a fruity character that may come from slow or arrested fermentation (e.g., keeving) or approximated by back-sweetening with juice. Notes of spicy-smoky, phenolic, and farmyard are common but not required (just as with, but subtler than, the English), and must not be pronounced. Ciders from Asturias typically have fresh citrus peel and floral aromas. Ciders from the Basque Country may also exhibit light spice, leather, and smoke aroma. Aged cheese and butter aromas and green olive juice notes may also be encountered, but any excess is undesirable. Acetic acid aromas are also common in these ciders.

Appearance. English-style and French-style may be barely cloudy to brilliant; medium yellow to amber color. Spanish-style ciders are unfiltered, so cloudiness is normal. Shaking the bottle before opening and pouring is recommended. The color for Asturian ciders should be straw yellow. The color for Basque ciders tends toward pale to deep gold. Tasting competitions can require specific visual evaluations after the traditional pouring:

> Espalme—foam must disappear quickly from the top of the cider.
> Aguante—small bubbles disappear slowly, allowing time to drink in perfect condition.
> Pegue—thin film favorably adheres to the sides of glass after cider has been drunk.

Mouthfeel. Medium body. Moderate to high levels of tannins for slight to moderate astringency. Bottle-fermented or bottle-conditioned ciders may have high carbonation, up to champagne levels, but not gushing or foaming. Pleasant scratchy and tickly throat due to acetic acid is expected (often more intense in Basque ciders).

Overall impression. English-style ciders are generally dry, full-bodied, and sometimes harsh. They will have a complex flavor profile and long finish. French-style ciders are typically made sweet to balance the tannin levels from the traditional apple varieties and are rich in body. Spanish-style ciders are dry, fresh with lively acidity.

Specialty

Hopped ciders

Made with the addition of hops.

Aroma/flavor. Cider and fresh apple character should be present and not completely overpowered by the hops. The choice of hops may influence fruit (e.g., citrus) and floral aromas.

Appearance. Clear to brilliant; hop addition generally does not contribute to color.

Mouthfeel. Moderate to full bodied. May be bitter from effect of hops, but not overly bitter.

Overall impression. A modern-style cider with added complex flavors. Lighter and fruitier alternative to India pale ales (IPAs).

Mixed fruit ciders

Made with the addition of other fruits or juices (non-apple).

Aroma/flavor. Cider character present, but not conflicted by the other fruits.

Appearance. Clear to brilliant; color reflective of added fruit, but should not show oxidation characteristics (e.g., red berries should provide red-to-purple color, not orange or brown).

Mouthfeel. Substantial. May be significantly tannic, depending on fruit added.

Overall impression. A modern cider with complex fruit flavors.

Spiced ciders

Also known as herb or vegetable ciders, include ciders made with culinary spices, herbs, and vegetables, as well as nuts, coffee, chocolate, spruce tips, rose hips, hibiscus, rhubarb, and the like. It does not include culinary fruit or grains. Flavorful fermentable sugars and syrups (agave nectar, maple syrup, molasses, treacle, honey, etc.) can be included only in combination with other allowable ingredients, and should not have a dominant character.

Aroma/flavor. Cider character present, but not completely overpowered by the spices, herbs, and/or vegetables.

Appearance. Clear to brilliant; color may be reflective of added ingredients.

Mouthfeel. Moderate to full bodied. Cider may be tannic from effect of added ingredients, but must not be bitter from over-extraction.

Overall impression. Like a modern cider with complex flavors. Apple character marries with the added fruits giving a balanced result.

Wood-aged ciders

Barrel-fermented or barrel-aged in which the wood and/or barrel character is a notable part of the overall flavor profile. Cubes, chips, spirals, staves, and other alternatives may be used in place of barrels.

Aroma/flavor. Cider character present, but not overpowered by wood/barrel character.

Appearance. Clear to brilliant; color that of a standard cider unless wood/barrel character is expected to contribute color.

Mouthfeel. Moderate body. May show astringent or heavy body as determined by wood/barrel character.

Unlimited ciders

Include products that may approximate a standard cider, but unlike a standard cider, these products will have carbonated water, malic acid, natural flavors, artificial flavors, and similar ingredients listed on the label. Generally, mass-market products that use modern production techniques such as high brix fermentation, chaptalization, and flavorings.

Distilled and Intensified

Ice ciders

A style that originated in Quebec in the 1990s. Juice is concentrated before fermentation, either by freezing the fruit before pressing it, or by freezing the juice and then removing water as it thaws. The fermentation stops or is arrested before the cider reaches dryness. No additives are permitted in this style; in particular, sweeteners may not be used to increase gravity. The character differs from a chaptalized cider (apple wine) in that the ice cider process increases not only the sugar (hence alcohol), but also the acidity and all fruit flavor components proportionately.

Aroma/flavor. Fruity, smooth, sweet-tart. Acidity at a level to prevent it being cloying.

Appearance. Brilliant; color deeper than a standard cider, gold to amber.

Mouthfeel. Full body. May be astringent and/or bitter, but this should be slight to moderate.

Fortified ciders

These are strengthened in alcohol (and aroma and flavor) after fermentation by the addition of spirits. A cider fortified with apple spirits is known in France as *pommeau*. Fortified Cider includes pommeau, pommeau-like products, and emulations of fortified wine styles such as port, sherry, or vermouth made with a base of apples and/or pears rather than grapes. A range of sweetness is possible by choosing how far into primary fermentation to add the spirits. Pommeau has high concentrations of residual sugar and is therefore relatively sweet and has retained fruity flavors. Contrarily, cider that has been allowed to ferment mostly or completely to dryness before the spirit addition will be much less fruity and is known to some as royal cider. Whether sweet or dry, fortified ciders are full-flavored and heavy-bodied, but not as intense as ice ciders. Fruit should be forward, acidity well-balanced, and spirits evident and warming but not harsh.

Spirits

These encompass distilled apple products. Unaged spirits are typically known as *eau de vie* and wood-aged spirits are typically known as brandy. Cider that is concentrated by freezing after fermentation (often known as applejack) are often excluded from competition due to adverse effects related to toxicity.

Eau de vie. White, non-oak-aged brandy. Usually ranges from 60 to about 100 proof, though products under 80 proof are uncommon in the U.S. Imported examples bear special labels. The spirit should be round in the mouth and free of heads (acetates and aldehydes—nail polish remover) or tails (fusel oils, often stemmy or fuel-like, they bead in an empty glass, and are most evident as smell in an emptied glass). Usually subtle on the fruit and may carry a hint of spice. Varietal character may be difficult to discern.

Brandy. In general, oak-aged brandies follow the same guidelines as eau de vie. Head and tail characters are faults. Apple is subtle. Barrel character may range from toasted wine barrel (toast and coconut, light yellow color) to charred whisky barrel (smoke, spice, dark amber color), but barrel character should be apparent. Apple brandies fall into two main camps: Calvados (France) and Applejack (Mid-Atlantic U.S.). Calvados is fermented from a French style cider and is dominated by the wild fermentation flavors. These accentuate the heads. They can give a green apple impression to some tasters, but commercial samples are often excellent candidates to illustrate acetaldehyde, ethyl acetate and (rarely) diacetyl. A calvados-style apple brandy should be heady, and should have muted barrel character. Applejacks should not evidence heads, but rounded, generic apple or apple blossom flavors and aroma, varietal character being a plus. Some spiciness is acceptable, partnered with smooth barrel character that should not overwhelm.

Origins of Cider Quality Attributes **5**

Cider quality is affected at each stage of the production process from branch to bottle. A wide range of apple cultivars and growing climates in addition to orchard management practices allows for countless variables in starting fruit material. Likewise, pre-fermentation treatments of fruit and juice, fermentation management, and post-fermentation treatments of cider will also have impacts on final product quality.

In the Orchard

Cultivar

Cider can be made from any apple, but cultivar selection will greatly impact the style and profile of the beverage. Apples can be classified, based on usage, as one of three types: "dessert," "culinary" (also referred to as "processing" or "cooking"), or "cider," and in some cases multiple types, "dual purpose." Dessert apple cultivars, such as "Fuji" and "Honeycrisp," are characterized by having colorful thin skins, juicy flesh, and mild to strong sweetness that provide for a pleasant raw consumption experience (Fig. 14). Culinary apple cultivars, such as "Granny Smith" and "Bramley's Seedling," are generally of large size and have a high acid content that aids in the retention of flavor and maintenance of shape when the fruit is subjected to heat. Cider apple cultivars, such as "Dabinett" and "Kingston Black," have relatively high phenolic levels that contribute to a bitter taste and/or astringent mouthfeel, which provide for an undesirable eating apple but an excellent fermented beverage. The analytical classification system established at LARS in Bristol, England (see Chapter 1, Table 1) further differentiates cider apples into bittersweets and bittersharps, based on juice tannic acid and malic acid levels. In the English classification system, the sharp classification can be taken to represent culinary and dessert apples. Sweets are very low in acid and unbalanced for fresh eating.

Fig. 14. Photographs of "Fuji", "Granny Smith," "Kingston Black," and "Bramley's Seedling" from the UK National Fruit Collection (Crown Copyright, 2019).

Moulton and King (2015) provides for a colorful guide to the utilization(s) of a broad selection of apples. Table 8 gives a sample of the guide (the whole guide is provided in Appendix C). Typically, a blend of apples is used to obtain a balance of sugar, acid, and tannin. Though not impossible, this balance is more difficult to achieve with the use of only one cultivar (See Appendices A–B). The cultivars used will depend largely on the intended style (see Chapter 4) and the availability of fruit. In the U.S., bittersweet and bittersharp apples are in short supply. As of 2016, dessert apple acreage in WA state, which accounted for 64% of national production, was reported to be a little over 179,000 (NASS, 2017), while cider apple acreage in the whole country was estimated (no official reporting) to be less than 500. The average price per pound was $1.62 for marketable dessert apples, $0.35 for cider apples, and $0.17 for non-marketable dessert apples (NWCA, 2016; NASS, 2017;), apples that do not meet fresh market standards but can provide for a juice base in fermentation.

Climate

A French term utilized to refer to the multiple environmental factors (e.g., climate, geographic location, soil composition) that influence the character

Table 8. What to do with your apples after they are picked? (modified from Moulton and King, 2015) (see also Appendix C).

Dessert (De)	—apple can be used for eating fresh out of hand
Culinary (Cu)	—apple for culinary uses; (P) good for pies, (S) good for sauce
Cider (Ci)	—apple can be used for making fermented beverage

Variety	De	Cu	Ci	Comments
Arkansas Black	●	P		Firm, moderately juicy
Ashmead's Kernel*	●		●	Russet skin, very firm, dense flesh
Bulmer's Norman*			●	Sweet, astringent, fast-fermenting juice, and mildly bittersweet cider
Cox's Orange Pippin*	●	P	●	Classic English dessert apple; pear-like aroma when baked, high in vitamin C
Fuji	●	S		Sweet, firm flesh (*early strains like Beni Shogun will ripen in western WA)
Gala	●			Rich sweet flavor, very juicy, dries well
Golden Delicious	●	S, P	●	Crisp, juicy flesh; very good for pies or baking whole, best flavor develops in cooking
Golden Russet*	●	S, P	●	Russet skin, sweet, crisp, fine-textured flesh, dries well
Granny Smith	●	S, P	●	Hard, firm, tart, develops mellow flavor when fully ripe after storage
Gravenstein*	●	S	●	Thin-skinned, juicy, sweet, short storage only; top rated for sauce, but not recommended for pie
Jonagold*	●	P, S	●	Very well balanced sweet-tart flavor; large fruit, holds up well in baking
McIntosh*	●	S	●	Highly aromatic, spicy, doesn't keep well
Red Delicious	●			Can reach very good quality if properly harvested and stored, not well adapted to Puget Sound
Spartan*	●			Very flavorful, firm white flesh; McIntosh cross
Hewe's Crab*			●	Juice ferments very slowly for highly flavored dry cider
Yarlington Mill*			●	Sweet, slightly astringent juice and a medium bittersweet cider

*Adapted to western WA conditions.

of a crop is *terroir*. This term is the basis of the French system (*appellation d'origine contrôlée*, AOC) for certifying geographical origins of wines through appellation (i.e., legal titles), as a means of regulation. Champagne is one of the most well-known appellations that can only be applied to wines originating from the historical province of Champagne in northeast France. In the U.S., there are similarly American Viticultural Areas (AVAs) that define geographical wine grape growing regions, as in turn defined by the Alcohol and Tobacco Tax and Trade Bureau (Title 27 Code of Federal Regulations Part 9). The concept of *terroir* is increasingly being utilized in the U.S. cider industry, and it is foreseeable that appellations could be pushed for in the major apple growing regions of Washington and New York, but it is not yet clear whether this strategy of diversification is backed by scientific data. The wine industry has been active for a longer period of time in the U.S. than the cider industry, and research on grapes and wine produced in the U.S. is much more extensive than on cider apples and cider. Research into the influence of the various soils and microclimates of the U.S. on cider apple quality is greatly needed. At the same time, such research requires a clear understanding of the specific quality attributes that are important to cider, which currently is not uniform among consumers, growers, and cidermakers in the U.S.

Wine grapes, which predominantly differ from their table grape counterparts in terms of phenolic content and aromatics, could potentially serve as a model for researching climate effects on cider apple quality. For example, wine grapes from cooler regions do not ripen as quickly, which can result in lower sugar content, higher acidity, and delicate tart fruit aromas such as cranberry and green apple. Growers in cooler climates face unique management issues that include potentially lower yielding vines and frost damaged or killed vines. In contrast, wine grapes from warmer regions ripen faster, accumulate more sugars, and, as a result, can have higher potential alcohol concentrations. Wines produced from warmer climate grapes have been described as possessing fuller bodies and flavors, dominated by fruity flavors including plum, blueberry, and blackberry, but may not retain as much acidity. Growers in warmer climates also face unique management challenges. Growers may struggle to retain acidity in the grapes, which generally declines as sugar accumulates, and to maintain the freshness of the wine. Grapes grown in hotter climates tend to have thicker skins, which, depending on the conditions and duration of extraction, can contribute more tannin. Alexander et al. (2016a) found no differences in the juice quality characteristics of four cider apple cultivars grown in northwest Washington, which has a cool, humid summer climate (16.0 °C on average) and central Washington, which has a hot, dry summer climate (22.1 °C on average). There was significant year-to-year variability in the quality characteristics measured, supporting the importance of testing juice quality every season. More investigation on climate impact on apple and cider flavors is needed for a more comprehensive understanding.

The influence of climate and soils on which apples can be grown where in the U.S. and to what degree of success, in terms of production factors such as yield, growth habit, and disease resistance, is better known. Regional cultivar performance data can be found in books such as *The*

Apples of New York (Beach, 2015 [1905]), and online via university cider research program websites including that of Michigan State University, Montana State University, University of Vermont, University of Wisconsin, and Washington State University. Apple trees in general can be grown in a range of soils, from medium textured clay to gravelly sands, but they perform best in soils that are well-drained, have a high organic matter content (decomposed plant and animal residues), and are nearly neutral in pH (5.8–6.5). Wet, low nutrient soils of extreme pH can restrict tree growth and provide for poor fruit development, which overall equates to low fruit quality if any. To achieve a consistent crop load of high quality, apple trees require exposure to a certain number of hours below a temperature threshold (generally 45 °F) over the winter (chill hours, CH) and a sufficient number of days of heat in the spring and summer (growing degree days, GDD). The CH and GDD requirements of apples vary with cultivar, some requiring as little as 200 CH and some more than 1,000 CH. Variability in performance due to climate is a large factor in why some cultivars are not uniformly praised by growers across regions and why it can be difficult to suggest which cultivars to plant for cider production. For example, "Ribston Pippin" was favored highly among apple growers in England but not New England, and "Kingston Black" is loved and hated across the western U.S. coastline.

Tree management

The relationship between orchard management and cider quality is much less understood than that of vineyard management and wine quality. However, as the cider industry continues to grow, this segment of research is also increasing. Alexander *et al.* (2016a) attributed differences in juice quality characteristics of multiple cider apple cultivars grown at the New York Agricultural Experiment Station, LARS, and WSU NWREC to differences in cultural practices across locations. Cultural practice differences for Washington and southwest England include water and nutrition management. The LARS data was obtained from a period (1905–1975) during which the studied cider cultivars were predominantly grown on standard-sized rootstocks in pastoral orchards. In contrast, the data obtained at WSU NWREC was collected from cider cultivars predominantly grown on dwarfing rootstocks that were relatively intensively managed. In northwest and central Washington, the orchards were drip or micro-sprinkler irrigated and foliarly or ground fertilized (predominantly nitrogen) to avoid deficiencies. The trees sampled for the LARS data were most likely more stressed than the trees sampled in the current study, and this difference could explain the lower tannin levels in Washington as greater stress has been correlated with greater fruit tannin content (Lea and Beech, 1978). With lower levels of tannin, Washington cidermakers may need to use a larger proportion of cider apples than is utilized by English cider makers to achieve the same level of bitterness and astringency. WSU researchers recently (April 2019) initiated a project to evaluate the use of regulated deficit irrigation in cider apple orchards as a means to produce a more desirable, complex fruit while conserving a critical natural resource.

Peck *et al.* (2016) examined the impact of fruit thinning on juice and cider quality and found that crop load density may impact the quality of juice and cider made from "York" apples. In particular, high crop loads were observed to decrease yeast assimilable nitrogen (YAN). YAN is important for yeast nutrition, though some cidermakers, particularly in the U.K. and France, may intentionally keep YAN very low to encourage slow fermentations. Fungicide applications in the orchard have in some cases been observed to have an effect on cider quality. Boudreau IV *et al.* (2017) found that the presence of the fungicide fenbuconazole resulted in a decreased fermentation rate and increased residual sugar at the end of fermentation, but that this could be alleviated with increased YAN supplementation. It was also observed that the presence of elemental sulfur, which is also applied as a fungicide, can increase the production of hydrogen sulfide which can impart a rotten egg aroma. Proper pruning and training of apple trees in general is important to achieving full skin color, maximum sugar accumulation, and manipulating fruit size. Neglected, crowded, shaded, damaged, and/or diseased wood can lead to fruit of poor quality as a result of decreased sunlight interception, increased competition for resources, and increased disease or pest pressure. As discussed by Forshey (1976), pruning represents a significant cost of apple production, and while the effects of inadequate pruning are not immediately evident, fruit quality is usually the first output to be impacted.

Fruit management

Several physiological and chemical changes occur during fruit maturation. Generally, as apples ripen, starch is converted into sugars by hydrolysis (Fig. 15), malic acid is degraded through respiration, and fruit firmness decreases. Apples harvested before full ripeness may be stored in modified or controlled atmosphere storage for long periods of time, allowing for transport across continents and marketing over multiple years. While post-harvest storage is commonly done for fresh-marketed dessert apples, apples used for cider production tend to be harvested closer to full ripeness for maximum conversion of starch into sugars and maximum production of volatile aromas (Girard and Lau, 1995; Lea, 2014; Alberti *et al.*, 2016).

In some regions, such as western Washington, it is a common practice among cider apple growers to store hand-harvested fruit at ambient temperature (e.g., in a barn) for up to two weeks to increase their sweetness and ease to juice (Fig. 16). However, if cider apples are being harvested

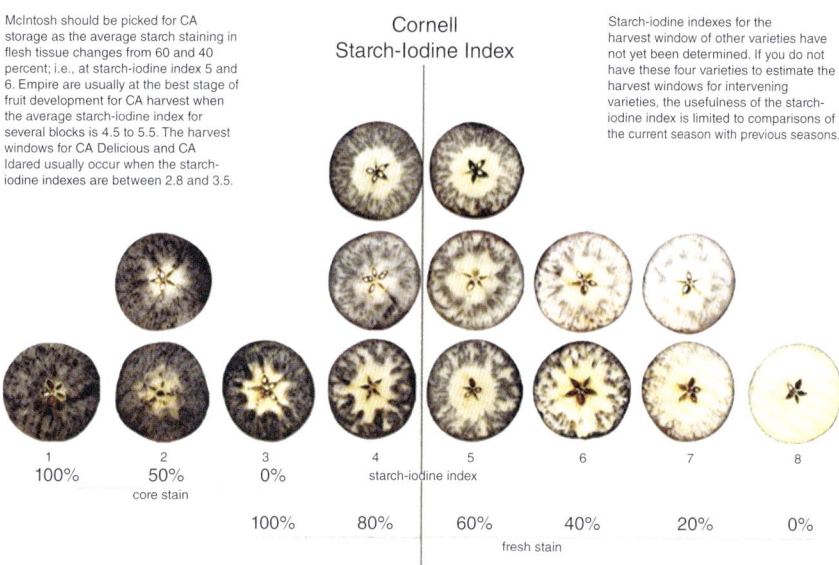

McIntosh should be picked for CA storage as the average starch staining in flesh tissue changes from 60 and 40 percent; i.e., at starch-iodine index 5 and 6. Empire are usually at the best stage of fruit development for CA harvest when the average starch-iodine index for several blocks is 4.5 to 5.5. The harvest windows for CA Delicious and CA Idared usually occur when the starch-iodine indexes are between 2.8 and 3.5.

Cornell
Starch-Iodine Index

Starch-iodine indexes for the harvest window of other varieties have not yet been determined. If you do not have these four varieties to estimate the harvest windows for intervening varieties, the usefulness of the starch-iodine index is limited to comparisons of the current season with previous seasons.

1	2	3	4	5	6	7	8
100%	50%	0%					
	core stain		starch-iodine index				
		100%	80%	60%	40%	20%	0%
				fresh stain			

Fig. 15. The Cornell starch-iodine test chart with a 1–8 ripeness scale, decreasing starch staining observed as apple ripens (Blanpied and Silsby, 1992).

Fig. 16. Hand-harvested "Brown Snout" cider apples (Photo courtesy of Northwest Cider Association).

fully ripe, the increase in sweetness is most likely not a result of an increased sugar content but rather a decreased water content (i.e., a concentration effect). The term for this storage process, "sweating," is derived from the dehydrated appearance of the fruit. Research on the enhancement of flavor development and ease of juice extraction in response to the practice of sweating is needed. Furthermore, if growers desire to mechanically harvest their cider apples, sweating of fruit is not recommended as significant yield losses due to rotting of bruised fruit has been shown by Alexander *et al.* (2016b). There has been increased research interest on how fruit maturity and storage impact cider quality factors, specifically in relation to tannins (Alberti *et al.*, 2016; Laaksonen *et al.*, 2017; Ewing *et al.*, 2019). In all of these studies, the effect on juice and cider quality was highly cultivar-dependent, and though tannin increases have been observed as a result of harvest date and increased storage times, results have not been consistent across cultivars, growing years, and growing regions to make definitive conclusions at this time.

In the Cidery

Pre-fermentation

Treatments include those that are applied to the fruit or juice prior to yeast inoculation.

Maceration. Crushed apple pomace can be immediately pressed (Fig. 17) or set aside for an amount of time before pressing to promote the extraction of phenolics and pigments from the skin and seeds, the latter known as maceration (Pinsley, 2019). However, it is important to note that time and temperature alone are unlikely to be sufficient for phenolic extraction, and enzymes and pulsed electric fields may be needed for significant results. Maceration does allow, however, for natural or added pectic enzyme action to aid in clarification. This is a step commonly employed during the keeving process to increase the clarification of the juice. Maceration is also used with especially high tannin ciders and perry fruit to increase oxidation of tannins and thereby decrease bitterness and astringency characteristics (Jolicoeur, 2013). Increased oxygen contact also results in browning reactions caused by the polyphenol oxidase (PPO) enzyme. Maceration durations can range from several minutes to several hours. However, prolonged maceration times, particularly at warmer temperatures, present a greater risk of spontaneous fermentation and contamination from indigenous yeast and bacteria.

Enzymes and clarification. Fresh-pressed apple juice is cloudy due to the presence of pectins and the colloids they form. Pre-fermentation clarification may be desired to obtain a clear product and improve the overall sensory qualities of the cider and to facilitate filtration. Reduction of solids prior to fermentation has its benefits. In grape wine, it has been observed that hydrogen sulfide odors may be reduced by removing solids that may have elemental sulfur residues, and the removal of esterase with the solids may help maintain greater levels of fruity ester aromas (Singleton *et al.*, 1975; Boulton *et al.*, 1999). In cider, however, reduction of juice solids will

Fig. 17. Apples must be crushed before pressing. The product of this process is called the "pomace" and may be set aside for a period of maceration prior to pressing (Photo courtesy B. Valliere, WSU).

often decrease the production of fusel alcohols, such as 2-phenylethanol, which is known for providing a "cidery" aroma (Beech, 1972). Pectic enzymes, or "pectinases," may be added to the ground pomace or pressed juice prior to fermentation to clarify cloudy juice. Pectinases break apart pectin chains and release the encapsulated contents. Compounds may remain in the juice solution, settle, or volatilize out of solution, particularly at warmer temperatures. While clarification may happen naturally or with the use of clarification aids post-fermentation, pectinase efficacy decreases with the presence of ethanol and will not be as effective at clarifying finished cider.

Thermal juice pasteurization and concentration. Thermal treatments to juice, including pre-fermentation pasteurization and concentration by thermal evaporation, may impact the quality of the juice and therefore the fermented cider. Thermal treatments to juice and concentrate may alter the color, aroma profiles, and nutritional content of juice (Su and Wiley, 1998). The type and extent of these changes will depend on the exact processes used but may significantly impact the organoleptic qualities of the resulting cider, particularly if significant nutrient degradation or chemical changes to aromatic precursors occur.

Fermentation

Flavors resulting from fermentation include ethanol, higher alcohols, esters, ethers, aldehydes, and acids. Secondary aromas are the most important for flavor development and are highly dependent on yeast strain and fermentation conditions (Pisarnitskii, 2001).

Yeast selection. Typically, Champagne or white wine *Saccharomyces cerevisiae* yeast strains are chosen to make cider. However, other wine or ale yeasts may also be selected depending on the desired outcome. Yeast strains may be selected based on their fermentation time, ethanol tolerance, or tendency to produce certain aromatic compounds, such as esters. Some yeast strains are also characterized as having low-nitrogen needs or the tendency to not produce sulfur-containing aromas. Ciders made with the same starting juice but different yeast strains can significantly differ from one another organoleptically due to the differing metabolisms and utilization of aroma precursors (Williams, 1974; Peng *et al.*, 2007).

Yeast nutrients. YAN, which is composed of primary amino nitrogen and ammonia, is an important nutrient for yeast metabolism. Insufficient nutrient concentrations may cause slow fermentations, stuck fermentations, or hydrogen sulfide off-aromas. Furthermore, variations in nutrient content and composition result in varying production of volatile aroma compounds. Individual yeast strains consume nutrients differently, resulting in even more variation in aroma compounds (Eleuterio dos Santos *et al.*, 2015). As YAN concentrations in apple juice can vary widely, and to prevent the formation of off-aromas or fermentation problems, exogenous nutrients may be added prior to and during fermentation. Nutrients are added in the form of diammonium phosphate, complex amino acid and vitamin blends, and oftentimes a mixture of the two. Yeast utilize different forms of nitrogen or metabolic functions, so nutrient composition may also have an effect on final cider aroma and quality (Bell and Henschke, 2005).

Fermentation temperature. The sensory properties of cider are also affected by temperature during fermentation. Fruity esters have been shown to be in higher concentrations in low temperature fermentations (10–15 °C) compared to high temperature fermentations (28–30 °C). This could be attributed to a change in yeast metabolic process and potentially evaporative losses at higher temperatures (Killian and Ough, 1979; Molina *et al.* 2007). Yeast strains typically are sold with a recommended temperature range for fermentation, and deviation from this range could lead to unsuccessful fermentations or unexpected aromatic production.

Post-fermentation

Treatments include those that are applied from after primary fermentation to bottling.

Malolactic fermentation. The bacterial conversion of malic acid to lactic acid is known as MLF (malolactic fermentation). Though not a true fermentation and therefore sometimes referred to as "malolactic conversion," MLF has been the standard term for decades and continues to be referred to as a secondary fermentation, as it typically follows alcoholic fermentation. This process is carried out by *Oenococcus oeni*, *Pediococcus*, and *Lactobacillus* species and may be encouraged by deliberate inoculation of bacteria (Wibowo *et al.* 1985). Bacteria may also be naturally occurring on the fruit or equipment. MLF may be desired when the final product is too sour and a reduction in acidity helps bring balance to the product. MLF may also result in an increased production of both positive and negative aroma compounds. In particular, diacetyl, a buttery aroma, is a common characteristic of ciders and wines that undergo MLF (Martineau *et al.*, 1995). Though MLF can be favorable to overall quality, it can also be detrimental if it is undesired or unmonitored. Ciders are at particular risk for this latter case, as essentially all apple and cider acidity can be attributed to malic acid. Therefore, while wine has a large proportion of tartaric acid that will not be affected during MLF, cider can potentially lose all of its acidity during uncontrolled MLF, resulting in a flat, unbalanced beverage. Furthermore, aroma compounds produced during MLF cannot always be predicted, particularly when MLF begins spontaneously without deliberate inoculation of a starter culture. Furthermore, aroma compounds produced during MLF cannot always be predicted, as research characterizing the aroma compounds produced by indigenous strains is still in its infancy (Sánchez *et al.*, 2014).

Enzymes and fining agents. For post-fermentation clarification, pectinase may once again be used, but as mentioned before, its efficacy in a high ethanol solution will be significantly less. Fining agents may also be used to aid in clarification and removal of other undesired characteristics. Both pectic enzymes and fining agents may also be used to aid in filterability by reducing solid content in the cider. There are many types of fining agents used in cider and wine for a wide variety of purposes. Though fining agents are often used for product clarification, some may also be used to remove quality attributes such as brown color, hydrogen sulfide off-aromas, excessive bitterness, and excessive astringency. Common fining agents include bentonite, protein fining agents (gelatin, casein, isinglass, egg albumen), carbon, silicon dioxide, alginates, and polyvinylpolypyrrolidone (Zoecklein *et al.*, 1990). Each fining agent may affect other aspects of the cider beyond its intended use, such as aroma or color, so cidermakers are encouraged to conduct fining trials of both the agent and the dose prior to addition.

Filtration. Filtration of cider may be employed to improve the clarity of the cider and reduce microbial loads. There are several types of filtering methods, and degree of clarity and removal of particulates will depend largely on the type of filtration method, the grade of the filter, the integrity of the filter, and successful operation of filtration. Filtration may aid in physically removing micro-organisms that could lead to spoilage problems (Bartowsky, 2009). Crossflow filtration, which efficiently filters cider through an automated tangential flow mechanism, has become more affordable and accessible for producers in recent years. This method of filtration has replaced many other traditional filtration systems and has decreased the use of some fining agents.

Blending. Most ciders are made of more than a single cultivar, and complementary cultivars are selected based on their ability to together create a balance of sugar, acid, and tannin. Blending may occur before or after the fermentation process and may also include the addition of exogenous sugars, acids, tannins, fruit, spices, hops, botanicals, or other components.

Pasteurization. Post-fermentation pasteurization, typically conducted in the package, is commonly employed as a means of reducing microbial populations and potential spoilage issues, including re-fermentation and exploding bottles. Pasteurization may affect overall flavor. Pasteurization of apple juice has been observed to significantly affect aroma, but research on pasteurization effects on white wine has shown no difference in quality (Malletroit *et al.*, 1990; Su and Wiley, 1998; Aguilar-Rosas *et al.*, 2007). Organoleptic changes after pasteurization have not yet been thoroughly researched in cider, and the absence of a general pasteurization protocol in the industry may contribute to varied results.

Carbonation. A tingling mouthfeel sensation is achieved through carbonation of cider. Carbon dioxide can be added through injection of carbon dioxide gas into a beverage under pressure or through a secondary fermentation. In the latter case, additional sugar is added to the packaged cider along with a secondary yeast strain prior to sealing the bottle. This causes the yeast to proceed with a second fermentation, creating carbon dioxide gas and ethanol in the process. The effect of carbonation has been described as "tingling," "irritant," and "fizziness," and may also increase the perception of sourness, though studies on this show varied results. (Hewson *et al.*, 2009; McMahon *et al.*, 2017). In sparkling wines it has been shown that carbonation levels from 2.0 g/L and 7.5 g/L have perceivably more "bite" than still or less-carbonated wines (McMahon *et al.*, 2017). In the U.S., products qualifying for the Hard Cider Tax Rate will contain less than 6.4 g/L CO_2, but higher carbonation levels outside of this tax class are possible (Alcohol and Tobacco Tax and Trade Bureau, 2015).

Maturation and aging. Cider may undergo a period of maturation, the period between fermentation and packaging, and aging, the period between packaging and consumption. The purpose of maturation and aging are to improve the product, and results will depend on the cider chemistry, storage and packaging material, temperature, time, and overall management practices (Boulton *et al.*, 1999). Cider may undergo maturation in vessels made from different types of materials. For neutral influence on the cider, stainless steel, glass, or plastic vessels may be chosen. Cider may also be matured in oak barrels or vats to impart woody, toasty flavors. It is common for older barrels to be used for cider, reducing the overall impact of oak color and flavor component leaching into the beverage (Buglass, 2010). Despite the lack of oak influence from these older barrels, they may harbor bacteria that initiate MLF which is desired by some cidermakers.

New barrels are costly and may overpower some more delicate aromas found in cider. Used barrels that previously contained other alcoholic beverages, such as whisky and rum, can also be used to impart additional flavor characteristics to ciders. In lieu of more expensive barrels, oak chips or other adjuncts may be added at a lower cost. Oak chip origin, toasting level, and maturation time are factors that will cause aroma variation in ciders (Fan *et al.*, 2006).

Pairing Cider with Food　　　　　　　6

Though the concept of food and beverage pairings may typically provoke images of wine carefully selected by sommeliers to accompany a formal meal, the art and science of selecting an appropriate combination of food and drink can be applied to beer, spirits, and, of course, cider (Fig. 18). The wide variety of cider styles, as described in Chapter 4, illustrates the versatility of the beverage category and its potential for widespread food pairing applications. The bitterness of tannins, sweetness of sugars, and sourness of acids found in cider allow for endless flavor pairings and interactions with food. Food and cider pairing presents additional opportunities for cider producers to convey information and promote recognition of their product through a culinary experience. Pairing may also encourage educational engagement with consumers through events and workshops. Furthermore, cider and food pairing increases opportunities for regional food and beverage producers to collaborate and create a regional food identity geared towards locals and gastrotourists alike (Barber *et al.*, 2008).

Although general guidelines for pairing foods and beverages can be useful (Fig. 19), understanding the mechanisms of sensory responses to the chemistries of different foods allows for a much greater ability to make informed selections. For example, while the common strategy of pairing red wines with red meats and white wines with white meats provides some direction for the general consumer, it does not account for the complex nuances present in either the food or the beverage. Therefore, cider and food combinations should as often as possible be curated based on the sensory characteristics of each substance and potential interactions that may occur when paired. Ideally, pairings result in an experience where both the cider and the food are improved by the combination. Pairings are also successful, but less so, if the pairing elevates the qualities of only either the cider or the food. Unsuccessful pairings detract from the positive characteristics of either or both the cider and the food.

Before arranging a cider and food pairing, it is important to understand how the variety of flavors may mingle to create an overall impression.

Fig. 18. Cider is a versatile option for pairing with a wide variety of foods including cheese, nuts, fruits, and many others (photo courtesy of Northwest Cider Association).

As discussed in Chapter 2, these interactions can include mixture suppression and enhancement. For example, bitterness can be masked by sweetness. This effect can be demonstrated by adding sugar to a cup of black coffee, and the phenomenon is deemed a mixture suppression (Lawless and Heymann, 2010). Saltiness also has a suppressive effect on bitterness. Sweetness generally masks acid, which is the desired effect when adding sugar to a batch of lemonade that is deemed too sour. However, relationships between tastes can become more complicated. Sweetness can have a suppressive effect on saltiness, but low concentrations of saltiness can actually enhance sweetness. This latter example is deemed a mixture enhancement or hyperadditive effect (Green *et al.*, 2010; Lawless and Heymann, 2010).

General Strategies

General pairing strategies for food and beverages, namely wine, have been developed and employed. Before the application of any of the strategies, it is important to consider the sensory profile of both the cider and the food. One should have a general understanding of the cider style and how that style indicates the level of sweetness, sourness, and bitterness. In addition, one should consider the sweetness, sourness, bitterness, and saltiness of the

Fig. 19. Graphic created by the Northwest Cider Association for Washington Cider Week, 2015.

food. Aromatic profiles, such as levels of oak or fresh fruit, should be also taken into account, as well as the overall intensity of both the cider and the food, so as to not overpower the other.

Highlighting similarities

Pairing by common components would involve matching based on a shared set of flavor characteristics and is a method used to celebrate the shared flavor identity of that food (Goldstein and Goldstein, 2006; Dornenburg and Page, 2009; Arnone and Simonetti-Bryan, 2012; Kim and Lecat, 2017). For example, a cider with high acidity could be paired with a food also high in acid, such as a salad dressed with vinaigrette, to highlight the fresh, acidic flavors. Ciders with residual sweetness may be paired with dishes containing some sweetness, perhaps from a sweet sauce or glaze. Dessert ciders that are very sweet, such as ice ciders, may be appropriate pairings with desserts. High tannin ciders may successfully accompany bitter foods such as arugula or dark chocolate. In addition, pairing by common aromas may be a successful strategy. For example, a cider with notable citrus aromas may be consumed with a citrus-containing dish, such as lemon chicken or citrus salad.

Opposites attract

Another method to pair ciders with food is to make selections based on contrasting flavor components in an effort to bring balance to the meal (Goldstein and Goldstein, 2006; Dornenburg and Page, 2009; Arnone and Simonetti-Bryan, 2012). In this case, the understanding of mixture suppression and enhancement is critical. Sweeter ciders may be paired with acidic foods that are initially perceived as too sour, and more acidic ciders may be paired with sweet foods that are initially perceived as too sweet. These pairings can result in a greater balance of flavors between the cider and the food. A similar balance may be achieved between sweetness and bitterness. Salty foods, in particular, benefit from successful contrasting cider pairings. For example, saltiness can be balanced by a cider with residual sweetness. Acidic ciders may also be paired with salty foods to provide a refreshing palate cleanser.

Cleansing the palate

Naturally, cider may be consumed with dishes that also contain substantial levels of fat and protein, such as meats and cheeses. Meats and cheeses have long been paired with wines due to the ability of the wine acidity to "cut" through the creamy, fat components of food and thereby cleanse the palate (Goldstein and Goldstein, 2006; Dornenburg and Page, 2009; Jackson, 2009; Arnone and Simonetti-Bryan, 2013). There is evidence that lighter white wines have more flexibility and success with cheese pairings compared to red wines (King and Cliff, 2005), and, by this logic, cider could be an even more ideal accompaniment to a wide variety of cheeses. Ciders, which like wine are also acidic by nature, have the ability to cleanse the palate and allow for further full enjoyment of the food.

   Tannin-rich ciders, much like red wines, may be choice accompaniments for protein-rich foods. Astringency, a tactile quality characterized as causing a rough, dry sensation in the mouth, is due to tannin interactions with salivary proteins (Lee and Lawless, 1991). Pairing highly tannic ciders with protein-rich dishes is a strategy to lessen harsh astringency from tannins.

Location, location, location

A common pairing mantra states that "if it grows together, it goes together." This is the idea that food and beverages that come from the same region contain complementary characteristics due to the sharing of a common terroir (Fig. 20). An example of this would be pairing Basque *sidra*, which is known for its dry, acetic characteristics, with Bacalao (salted cod stew). Similarly, ciders from Normandy might best be paired with Camembert or

Fig. 20. Pairing by common terroir in the Pacific Northwest might mean having grilled salmon with some locally produced cider (photo courtesy of Northwest Cider Association).

lamb. For modern ciders, the tradition of cider and food may not be as historically linked, but development of connections and partnerships between cider and regional agriculture is another opportunity for newer cider regions to reach consumers.

Pairing outside the comfort zone

Though many food and beverage pairing suggestions focus on traditionally western foods, the increasing cultural diversity in North America and expansion of the global market provide ample ways to expand cider and food pairings to include a larger variety of culinary dishes. Cider is in a unique position to become the choice pairing for foods that are challenging to match with wine due to its relatively high alcohol content. Foods containing hot spicy flavors, such as those in chili peppers or curries, cause a chemesthetic warming and pain sensation (Green, 1996). High ethanol concentrations have been observed to increase the perception of heat (Demiglio and Pickering, 2008), making high alcohol wines incompatible for many consumers. However, as exposure to cold can counteract the burning sensation by inhibiting the receptor responsible (Babes *et al.*, 2002), chilled white wines have long been suggested as complements to Thai, Indian, and other cuisines regarded for their spicy heat (Simon, 1997; Goldstein and Goldstein, 2006; Jackson, 2009). Therefore, the relatively low alcohol concentration of ciders and tendency to serve cider chilled provides a perfect pairing opportunity for spicy, hot dishes. Recently, Kim and Lecat (2017) examined the potential for developing wine pairing criteria for Korean food and concluded that spicy dishes were best complemented with low-tannin, low-alcohol, mildly sweet, acidic wines to refresh the palate. Furthermore, sparkling wines with a relatively low price point were identified to be an attractive option to pair with Korean snack foods. Again, cider, a naturally low alcohol, often effervescent, acidic product with varying levels of sweetness and tannins could be an ideal pairing option for this food category.

The cider market continues to increase in North American and international markets, and opportunities to pair cider with foods that are representative of both the national and global multicultural population could help sustain the growing consumption of the beverage. Pairing cider and food is a subjective endeavor, and personal preferences regarding how much someone likes a cider or a food item will contribute to how much a person likes the pairing. For example, a person who has a strong liking for modern ciders may have a tendency to enjoy most pairings with that cider style compared to pairings with other styles. Culture and upbringing are also influential on an individual's preference, and genetics also influence an individual's ability to taste and smell certain substances (Jackson, 2009).

Cooking with Cider

Besides food pairings, cider can be harmoniously used as an ingredient in cooking. Now more than ever, recipes using cider can be found in cider magazines, newsletters, and blogs. Carr *et al.* (2018) recently released the *Ciderhouse Cookbook* which contains 127 recipes for utilizing cider apples in their raw and fermented forms, from incorporation in salads and soups to delivery of decadent desserts. *CIDERCRAFT* magazine (cidercraftmag. com) also contains a section on its website dedicated to cider-based recipes, including main dishes, desserts, and cocktails.

Cider can be used as a base for braising meats, poaching fruits for desserts, or adding a fresh acidity to salad dressings. The basic taste components of cider, namely sweetness, sourness, and bitterness, can complete dishes lacking in those particular aspects. Using cider as an ingredient in protein-rich dishes can also encourage the production of Maillard reaction products (i.e., non-enzymatic browning) when the cider contains some residual sweetness. The Maillard reaction, which is important for organoleptic quality development in food products such as coffee and baked goods, occurs between amino acids and reducing sugars during the application of heat. The result is browning and development of flavor compounds that produce aromas that include characterizations such as "toasty," "caramel-like," and "meaty" (Van Boekel, 2006). Therefore, the addition of cider containing residual sugars to protein-based cooking may encourage the development of these flavors.

Apple cider vinegar can also be used in cooking. The *Cooking with Apple Cider Vinegar Cookbook* by Martha Stephenson (2018) provides 40 recipes for "getting your daily dose of apple cider vinegar."

Appendix

A. Background Information on Varietal Data Collected at WSU NWREC

Each year from 2002 to 2015, two to nine harvested cider apple cultivars were selected to make a varietal cider, based on amount of juice available and program capacity. Fruit of each cultivar was harvested when fully ripe, utilizing a combination of harvest metrics (black seed coloring, minimum degree Brix reading of 12.0, and an 8 out of 9 starch index value [BC/Ontario starch index] (Blanpied and Silsby, 1992; Chu and Wilson, 2000)). From 2002 to 2012, fruit was shredded and pressed with a Correll Cider Press (Standard Large with 5 gal. capacity and 0.3 horsepower, CIDER PRESS LLC, Veneta, OR). From 2013 to 2015, fruit was shredded with a fruit mill (MuliMax 30 with a 2.8 horsepower, Zambelli Enotech, Camisano Vicentino, Italy) and pressed with a water press (Carezza with 10.5 gal. capacity, Enotechnica Pillan, Camisano Vicentino, Italy).

A standard fermentation protocol (Moulton *et al.*, 2010) was followed from 2002 to 2015 to produce varietal ciders. Key steps of the protocol are as follows: pH was measured with a pH meter (Orion 3 Star, Thermo Fisher Scientific, Pittsburgh, PA) and samples adjusted to be within a desired pH range of 3.3 to 3.7 if necessary (e.g., some samples requiring malic acid addition to drop the pH from 3.9). In choosing a range rather than targeting a specific pH for all the varietals produced in this study, there is the potential for undesired experimental variability as pH not only impacts the perception of acids but also of various aromatics that are released via the pH dependent activity of enzymes (Bakker and Clarke, 2011). Sulfite (potassium metabisulfite, $K_2S_2O_5$, Davison Winery Supplies, McMinnville, OR), serving as an antimicrobial, was added at a dosage of 150 mg L^{-1}. Yeast (Lallemand DV-10, Davison Winery Supplies, McMinnville, OR) was hydrated (100 g kg^{-1}, 40 °C), allowed to cool to 27 °C, and added to the juice at a dosage of 31 g hL^{-1}. Pectinase (100 g kg^{-1}; KS, Scott Laboratories, Petaluma, CA) was added at a dosage of 8 mL hL^{-1}. An initial specific gravity

(SG) reading was recorded utilizing a hydrometer (range 1.000–1.070, calibrated at 15.6 °C; VeeGee Scientific, Kirkland, WA). Juice was racked into 5.0 gal. carboys (ULINE, Pleasant Prairie, WI) with an airlock (Northwest Cider Supply, Monmouth, OR). Nutrients (Fermaid K, Scott Laboratories, Petaluma, CA), supplements for the yeast, were added at a dosage of 25 g hL^{-1}. When SG dropped below 1.025, approximately 72 h after nutrient supplementation, the carboys were topped off with extra juice. Fermentation was allowed to complete at 14 °C until SG was less than 1.000, approximately two and a half weeks after being topped off. The cider was left to mature at 14 °C. Sharp and sweet cultivars matured for an average of three months, and bittersweet and bittersharp cultivars matured for an average of six months. Multiple rackings were performed after reaching full maturity to clarify, with the final racking into a Cornelius keg (Home Brew Supply, San Marcos, TX). Sulfite was adjusted as needed (target range 30–40 parts per million [ppm] free SO_2, 0.8 ppm molecular SO_2). The final product was bottled in 12 and 25 oz bottles (Home Brew Supply, San Marcos, TX), pressurized and carbonated (55–69 kPa; Central Welding, Burlington, WA) at 14 °C, and capped. The finished bottles of cider were stored at 14 °C until they were evaluated.

From 2003 to 2015, each varietal was evaluated once by a minimum of 4 cidermakers. Optimally, varietals should have been evaluated by a minimum of 10 cidermakers over multiple years to minimize variation inherent to sensory perception and due to vintage. Financial and logistical factors limited WSU researchers' ability to acquire the optimal panel size. Evaluators were drawn annually from a pool of 10 regional well-respected cidermakers. At each evaluation, participants were re-familiarized with a standard cider tasting sheet (Fig. 10). Cider evaluations were conducted blind with high acid and low tannin samples presented first (Miles *et al.*, 2017).

B. Table of Varietal Data

Data produced from cider and dessert apples at WSU Mount Vernon NWREC, including acidity (Ac), bitterness (Bit), sweetness (Sw), astringency (Astr), and body (Bdy), where L = low, M = medium, and H = high (Moulton *et al.*, 2010).

Variety	Description	Color	Aroma	Ac	Bit	Sw	Astr	Bdy	Flavor
Amere de Berthcourt	Astringent with some aftertaste, good blender to boost bitterness	Gold	Confectionery, estery, honey, aspirin, grass, vanilla	L	MH	L	M-H	M	Honey, vanilla, Jolly Rancher, long bitter aftertaste
Braeburn*	Cider thin, not complex, would benefit from blending	Straw	Apple skins, confectionery, pear	L-M	L	L	L	L	Somewhat balanced but slightly acidic, lacks aftertaste, bland
Brown Snout	Good stand alone single-varietal cider; good body, balance and mouthfeel; juice will need acid adjustment prior to fermentation. Makes a darker viscous cider. Fruit bitterness is on the low-end, but cider is easy to drink even as a single varietal	Amber	Apple, nutty, banana, maple, caramel, resinous	M	M	L	M	H	Apple, slight banana, nutty, butterscotch
Brown's Apple	Good balance, a little bland, would benefit from blending, nice nose	Amber	Citrus, banana, butter, rummy	M	M	L	M	M	Citrus, apple, fresh cut grass, butterscotch, honey
Bulmer's Norman	Good balance, somewhat lighter body for a bittersweet, a little thin	Gold	Apple, honey, confectionery, pear drops	M	L-M	M	L-M	M	Cider a little thin, aftertaste short-lived but astringent, builds and lingers, refreshing

Continued

Variety	Description	Color	Aroma	Ac	Bit	Sw	Astr	Bdy	Flavor
Cap O' Liberty	Bittersharp tending to sour, unbalanced, bitterness lingers, body viscous but not heavy	Gold to slightly amber	Apple, bittersweet spicy, butterscotch, citrus, cidery	M-H	M-H	L	M	M-H	Long bitter aftertaste, sour, need to have acids balanced, complex spicy flavors that will be enhanced when balanced
Dabinett	Has body and astringency, tannins are harsh and tend to dominate flavor. Good alone for those who like a stout cider, otherwise blend to soften tannin; good nose	Amber	Apple, banana	M	M-H	L	H	M-H	Apple, raisins
Foxwhelp	No outstanding character, will add some body and lots of acid to blends. Best use as blender to give more acidity and body to a cider; earthy tones	Amber	Slightly spicy, banana, nutty	H	L-M	L	L-M	M	Apple, pear, sharp apple taste, butterscotch
Frequin Rouge	Viscous with some body, mildly bitter and astringent with some complex aromas. Makes a complex single varietal but may be best as a blender to boost tannins and add complexity	Amber golden	Cidery, nutmeg, bandaid, clove, vanilla, honey	L-M	M-H	L	M-H	M	Cidery, nutmeg vanilla, slightly warms the back of the mouth, some tartness but moderate bitter aftertaste lingers

Gala*	Cider thin but not watery, some fruit aroma, would benefit from blending for more body	Straw	Appley, pear drops, fruit salad, lemon	L-M	L	L-M	L	L	Thin, bland, short after-taste, not complex but interesting blend or aromas
Graven-stein*	Good aroma but cider not complex, sour with a lingering tang, fruit should be fully ripe	Gold	Apple sauce, apple, pear drops	M	L	L	L	L-M	Short after-taste, tang lingers, not complex but has refreshing character, easy to drink, sweetening would enhance balance
Harry Masters' Jersey	Good body and balance, good stand-alone varietal, fruity aroma	Amber	Woody, apple, winelike, resinous, floral, pineapple, perfume	M	M	L	M	M	Melon, berries, butterscotch, pine needles, tart, nutty
Jonagold*	Adds fruit both in aroma and flavor – best if used to bring fruitiness to a cider apple with tannins and body	Straw	Apple	M	L	L	L	L	Citrus, apple
Kermerrien	Cidery, estery; fruit flavored candy, creamy aftertaste, bitterness moderate to high but not harsh, good blender	Straw to gold	Cidery, spicy, bubblegum, estery, nutmeg, banana, bandaid, woody, honey	L-M	M-H	L	M-H	M	Fruity, banana, bubblegum, creamy, bitter aftertaste, that lingers but is not harsh

Continued

Variety	Description	Color	Aroma	Ac	Bit	Sw	Astr	Bdy	Flavor
Kingston Black	Stand alone single-varietal cider, tannins are soft, good balance, wonderful apple flavor jumps all over the mouth	Gold to amber	Apple, spicy, nutty	M	M	L	M	M	Citrus, apple, butterscotch
McIntosh*	Brings aroma to a cider—great for blending to enhance the bouquet, but not good as a single-varietal cider, lacks body and tannin	Straw	Winelike, apple, tropical, berry, floral	M-H	L	L	L	L	Winelike, apple, fruity, toast
Melrose*	Thin cider, uncomplicated, unpretentious, balance fair, slightly acidic	Straw	Butterscotch, grass, pear, cidery, cooked banana	M	L	L	L	L	Short aftertaste with slight citrus flavor
Michelin	This cider can stand alone but would benefit from blending. Good strong apple/berry aroma, tastes like cooked apple, bland	Amber golden	Apple, banana, nutty, berry, gooseberry, floral esters, caramelized apple	M	M	L	M	M	Fruity, apple, estery, berry, baked apple
Muscadet de Dieppe	A lot of body, good balance. Very good single-varietal cider, even when fermented to dryness still gives a slightly sweet taste, mix of fruit and grass flavors	Dark amber	Apple, banana nutty, tropical grassy	M	M	L-M	M	H	Fruity, apple, spice, nutty, honey, syrup, fresh cut grass

Variety	Description	Color	Aroma	Ratings	Tasting notes
Pinova* **(Corail, Sonata, Pinata)**	Cider thin, balanced, not complex, lacks body	Gold	Apple skins, confectionery, sweets, pineapple	M L L L L L	Short after-taste with citrus flavor
Reine des Pommes	Good balanced fruity cider with higher tannins, may be good single varietal stout type but when blended should build complexity, boost tannins, full bodymouth feel (astringency)	Amber golden	Apple skins, rosin, spice, citrus, lychee fruit, honey, tropical candied fruit, Jolly Rancher	L-M M-H L M-H M	Apple skins, spice, Jolly Rancher, citrus, full body that lingers, creamy, good fruit character, bitterness lingers and builds
Tom Putt	Mildly sharp cider, creamy mouth feel, balanced	Straw	Apple skins, bruised apple, lemon, grassy	L L M L L-M	Slightly sweet, lightly creamy, thin aftertaste medium that tapers off quickly
Vilberie	Very bitter with high astringency. Bitterness harsh; best used in blending. Very stout cider	Amber	Apple, spicy, floral	M H L H M	Bitterness dominated, estery, metal taste, tea bags, aftertaste copper penny, woody bark
WSU AxP Crab	Great blender, provides a lot of mouth feel because of its astringency. Some like it as a single-varietal cider but probably best for blending	Straw	Apple, spicy, nutty, tropical, floral	H M-H L H M	Fruity, winelike, fresh cut grass, citrus, vanilla

*Indicates dessert apple.

C. Guide to the Utilization(s) of a Broad Selection of Apples (modified from Moulton and King, 2015)

Dessert (De) —apple can be used for eating fresh out of hand
Culinary (Cu) —apple for culinary uses; **(P)** good for pies, **(S)** good for sauce
Cider (Ci) —apple can be used for making fermented beverage

Variety	De	Cu	Ci	Comments
Arkansas Black	●	P		Firm, moderately juicy
Ashmead's Kernel*			●	Russet skin, very firm, dense flesh
Ashton Brown Jersey			●	Full-bodied, bittersweet cider
Baldwin		P	●	Very sweet; good for pie when skin slightly green
Braeburn*	●	P		Tangy-sweet, very firm; great for whole baked apples
Bramley's Seedling*		S, P		Classic English culinary apple, high acid
Bulmer's Norman*			●	Sweet, astringent, fast-fermenting juice and mildly bittersweet cider
Calville Blanc d'Hiver	●	P		Classic French dessert apple, very high in vitamin C, tart, juicy
Centennial Crab*	●	S		Edible crabapple; good for jelly and pickling
Chisel Jersey*			●	Bittersweet and astringent juice, and a full bittersweet cider
Cortland, Redcort*	●	S, P	●	Sweet and acidic, very juicy; good for salads as it discolors very slowly after being cut
Cox's Orange Pippin*	●	P		Classic English dessert apple; pear-like aroma when baked, high in vitamin C
Elstar*	●	S, P		Tart flavor that mellows after storage; very good for baking whole
Empire*	●	S, P	●	Flesh white, sweet, crisp, juicy, and firm
Esopus Spitzenberg*	●	S, P		Very firm, crisp flesh
Foxwhelp*			●	Bittersharp that produces aromatic, musky flavored cider
Fuji	●	S		Sweet, firm flesh (*early strains like Beni Shogun will ripen in western WA)
Gala	●			Rich sweet flavor, very juicy, dries well
Golden Delicious	●	S, P	●	Crisp, juicy flesh; very good for pies or baking whole, best flavor develops in cooking

Continued

Variety	De	Cu	Ci	Comments
Golden Russet*	■	S, P	■	Russet skin, sweet, crisp, fine-textured flesh, dries well
Granny Smith	■	S, P		Hard, firm, tart, develops mellow flavor when fully ripe after storage
Gravenstein*	■	S	■	Thin-skinned, juicy, sweet, short storage only; Top rated for sauce, but not recommended for pie
Grimes Golden*		S		Very high sugar content
Hyslop Crab		S	■	Too astringent for fresh eating; excellent for jelly and pickling, dries out quickly so use immediately; blend for cider
Idared*	■	P		Thick skinned, highly aromatic, slightly tart, juicy
Jonagold*	■	P, S		Very well balanced sweet-tart flavor; large fruit, holds up well in baking
Jonathan	■	P	■	Small, slightly tart
Lady (Pomme d'Api)	■			Very small and colorful, tender, crisp, juicy flesh; used for decoration
McIntosh*	■	S	■	Highly aromatic, spicy, doesn't keep well
Michelin*	■		■	Sweet, mildly astringent juice and medium bittersweet cider
Mutsu	■	S		Firm, crisp, sweet-tart at harvest, very sweet after storage; commercial name "Crispin"
Newtown Pippin (Albemarle Pippin)	■	S, P	■	Crisp, tender, sweet-tart flesh; very good for pies, not for salad as it browns quickly
Nonpareil	■			Sweet-tart flesh; Good varietal
Northern Spy	■	P	■	Juicy flesh high in vitamin C; blends with crab apples for cider
Northwest Greening*	■	P, S		Tough skinned, flesh firm, tart, juicy
Pitmaston Pineapple	■			Russet skin, small and very sweet, flavor likened to pineapple
Red Delicious	■			Can reach very good quality if properly harvested and stored, not well adapted to Puget Sound
Rhode Island Greening*	■	P		Tart greenish flesh; one of best for pies, also dries well
Rome Beauty		S	■	Very thick skin, crisp, tart flesh; excellent for baking whole as holds shape and has marvelous texture, not recommended for pies

Continued

Variety	De	Cu	Ci	Comments
Roxbury Russet*	■	S	■	Russet skin, very high sugar content, medium-acid fruit; fine clear cider with aromatic flavor
Smith's Cider*			■	Origin Bucks Co., PA, still cultivated for cider
Spartan*	■			Very flavorful, firm white flesh; McIntosh cross
Stayman Winesap	■	S, P	■	Juicy, firm flesh; Seedling of Winesap
Swaar*	■			Dense, firm flesh, high sugar and acid, mellows in storage
Sweet Coppin*			■	Sweet juice with no astringency and a sweet to mildly bittersweet cider
Virginia (Hewe's) Crab*			■	Juice ferments very slowly for highly flavored dry cider
Winesap	■	S, P	■	Very juicy flesh, strongly sweet-sour flavor; Good when baked whole; best blended in cider with bland, sweet apples
Yarlington Mill*			■	Sweet, slightly astringent juice and a medium bittersweet cider
York Imperial		S, P		Characterized by its superb keeping quality

*Adapted to western WA conditions.

Glossary

Airlock. Device that allows for the release of carbon dioxide, but excludes the entrance of oxidizing air or foreign particles. Functions essentially as a one-way valve.

Alcohol by volume (ABV). A standard measure of alcohol content (ethanol, C_2H_5OH) in a beverage (commonly reported as volume of alcohol as percent of total volume).

Burette. Graduated vessel for dispensing of known volumes of liquid, commonly available in plastic or glass.

Carboy. Fermentation vessel, commonly available in glass or plastic.

Color indicator. Compound that in solution changes color in response to pH, generally a weak acid or base.

Cornelius keg ("Corny" keg). Stainless steel cylinder that can be pressurized to a maximum of 130 PSI, pressurization (usually carbon dioxide, CO_2) moves fermented liquid from the bottom of the keg to a bottle or tap.

Endpoint (acid/base titration). The pH at which the chosen indicator changes color; the indicator ideally is chosen such that the endpoint is close to the equivalence point.

Equivalence point (acid/base titration). pH at which concentration of added titrant is sufficient to neutralize compound being quantified.

Fermentation. The chemical breakdown of a carbohydrate (e.g., glucose) into an alcohol (e.g., ethanol) or acid (e.g., lactic acid) by bacteria, yeasts, or other microorganisms.

Hydrometer. Device that measures the ratio of the density of a sample liquid to the density of water at a manufactured calibrated temperature, commonly used for measuring soluble solids or ethanol concentrations.

Lees. Sediment consisting generally of dead yeast cells and other solid matter (i.e., fruit particles, fining agents, etc.).

Oxidizing agent. Reagent that readily donates an oxygen atom or removes one or more electrons.

Parts per million (ppm). Unit of concentration often used to measure compounds in air or water; 1 ppm is one part in 1,000,000. The metric unit mg/L is equal to 1 ppm (parts per million).

pH. Concentration of hydronium ions on a negative logarithmic scale

Racking. Process of transferring liquid (juice or cider) from one vessel to another to separate it from the settled lees.

Redox indicator. Compound that in solution changes color at a specific electrode potential, voltage.

Reducing agent. Removes an oxygen atom or donates electrons.

Revenue. Amount of money generated from sale of goods, services, or use of assets before expenses are deducted.

Solubilized. Dissolution of a compound into a solvent, commonly water or ethanol.

Specific gravity (SG). Relative density of a liquid compared to the density of water at a set temperature. Density of water at 68 °F is 0.9999.

Strong acid. Acid which fully dissociates in water (HA <--> H^+ + A^-).

Strong base. Base which fully dissociates in water (BOH <--> B^+ + OH^-).

Sulfite. Compound that contains a sulfite ion, sometimes used as a preservative and fumigant for its antimicrobial properties.

Titrimetry. Method of quantifying the concentration of a compound by adding a reagent of known concentration.

Varietal cider. Cider produced from a single apple cultivar.

Viscosity. Internal resistance to flow; a measure of fluid friction.

Wavelength. Spatial measurement of electromagnetic radiation in nanometers.

Yeast. Single-celled microorganisms of the Fungus kingdom that are responsible for fermentation.

27 CFR 4.39 - Prohibited practices (April 1, 2017).

(a) Statements on labels. Containers of wine, or any label on such containers, or any individual covering, carton, or other wrapper of such container, or any written, printed, graphic, or other matter accompanying such container to the consumer shall not contain:

(1) Any statement that is false or untrue in any particular, or that, irrespective of falsity, directly, or by ambiguity, omission, or inference, or by the addition of irrelevant, scientific or technical matter, tends to create a misleading impression.

References

Aguilar-Rosas, S.F., M.L. Ballinas-Casarrubias, G.V. Nevarez Moorillon, O. Martin-Belloso, and E. Ortega-Rivas. 2007. Thermal and pulsed electric fields pasteurization of apple juice: effects on physicochemical properties and flavour compounds. *Journal of Food Engineering*, 83: 41–46.

Alberti, A., T.P.M dos Santos, A.A.F. Zielinski, *et al*. 2016. Impact on chemical profile in apple juice and cider made from unripe, ripe, and senescent dessert varieties. *LWT-Food Science and Technology*, 65: 43–443.

Alcohol and Tobacco Tax and Trade Bureau. 2015. *Code of Federal Regulations*, 27 CFR Part 24. U.S. Dept. Treasury, Washington DC, 17 March 2016.

Alcohol and Tobacco Tax and Trade Bureau. 2017. *Cider Statistics CY 2007–2016*. U.S. Dept. Treasury, Washington, DC.

Alexander, T.R., J. King, A. Zimmerman and C.A. Miles. 2016a. Regional variation in juice quality characteristics of four cider apple (*Malus xdomestica* Borkh.) cultivars in northwest and central Washington. *HortScience*, 51: 1498–1502.

Alexander, T.R., J. King, E. Scheenstra and C.A. Miles. 2016b. Yield, fruit damage, yield loss and juice quality characteristics of machine and hand-harvested "Brown Snout" specialty cider apple stored at ambient conditions in northwest Washington. *HortTechnology*, 26: 614–619.

Alexander, T.R, C.F. Ross, E.A. Walsh and C.A. Miles. 2018. Sensory comparison of ciders produced from machine- and hand-harvested "Brown Snout" specialty cider apples stored at ambient conditions in Northwest Washington. *HortTechnology*, 28: 35–43.

Alwood, W.B. 1903. A study of cider making. *U.S. Dept. Agr., Bur. Chem. Bul.*, 71.

Arnone, K. and J. Simonetti-Bryan. 2013. *Pairing with the Masters: A Definitive Guide to Food and Wine*. Delmar, Cengage Learning, Clifton Park, NY.

Babes, A., B. Amuzescu, U. Krause, A. Scholz, M.-L. Flonta and G. Reid. 2002. Cooling inhibits capsaicin-induced currents in cultured rat dorsal root ganglion neurones. *Neuroscience Letters*, 317: 131–134.

Bakker, J. and R.J. Clarke. 2011. *Wine Flavor Chemistry*, Chapter 4: Volatile components. Wiley-Blackwell, Hoboken, NJ.

Barber, N.A., J.R. Donovan, and T.H. Dodd. 2008. Differences in tourism marketing strategies between wineries based on size or location. *Journal of Travel & Tourism Marketing*, 25: 43–57.

Barker, B.T.P. 1903. *Classification of Cider Apples*. Natl. Fruit Cider Inst., Long Ashton Res. Sta., Bristol, England.

Bartowsky, E.J. 2009. Bacterial spoilage of wine and approaches to minimize it. *Letters in Applied Microbiology*, 48: 149–156.

Beach, S.A. 2015 [1905]. *The Apples of New York, Vol. 1* (Classic Reprint) Forgotten Books, London.

Beech, F.W. 1972. Cider making and cider research: A review. *Journal of the Institute of Brewing*, 78: 477–491.

Beech, F.W. and J.G. Carr. 1977. Cider and perry. *Economic Microbiology*, Vol 1., *Alcoholic Beverages* (A.H. Rose, ed.) pp. 139–313. Academic Press, London.

Bell, S.-J. and P.A. Henschke. 2005. Implications of nitrogen nutrition for grapes, fermentation and wine. *Australian Journal of Grape and Wine Research*, 11: 242–295.

Bender, D.A. 2009. *A Dictionary of Food and Nutrition*, 3rd edn. Oxford University Press, Oxford.

Blanpied, G.D. and K.J. Silsby. 1992. Predicting harvest date windows for apples. *Cornell Coop. Ext. Publ. Info. Bul.* 221.

Bore, J.M. and J. Fleckinger. 1997. Pommiers à cidre: Variétés de France. Quae, INRA Editions, Paris, Feb. 22, 2017.

Boudreau IV, T.F., G.M. Peck, S.F. O'Keefe, and A.C. Stewart. 2017. The interactive effect of fungicide residues and yeast assimilable nitrogen on fermentation kinetics and hydrogen sulfide production during cider fermentation. *Journal of the Science of Food and Agriculture*, 97: 693–704.

Boulton, R.B., V.L. Singleton, L.F. Bisson, and R.E. Kunkee. 1999. *Principles and Practices of Winemaking*. Chapman & Hall, New York, N.Y.

Buglass, A.J. 2010. *Handbook of Alcoholic Beverages: Technical, Analytical and Nutritional Aspects*, I and II. John Wiley & Sons, Ltd, Chichester, U.K.

Carr, J., N. Blum, and A. Blum. 2018. *Ciderhouse Cookbook*. Storey Publishing, North Adams, MA.

Chu, C.L.G. and K.R. Wilson. 2000. Evaluating maturity of "McIntosh" and "Red Delicious" apples. Ontario Ministry Agr. Food Rural Affairs Publ. Order No. 00-025. http://www.omafra.gov.on.ca/english/crops/facts/00-025.htm, accessed May 3, 2016.

Crews, E. 2007. Drinking in colonial America. *Colonial Williamsburg Journal*, Williamsburg, VA.

Demiglio, P. and G.J. Pickering. 2008. The influence of ethanol and pH on the taste and mouthfeel sensations elicited by red wine. *J. Food Agric. Environ.*, 6: 143–150.

Dornenburg, A. and K. Page. 2009. *What to Drink With What You Eat*. Bulfinch Press, New York, NY.

Eleutério dos Santos, C.M., G.A.M. Pietrowski, C.M. Braga, *et al.* 2015. Apple amino-acid profile and yeast strains in the formation of fusel alcohols and esters in cider production. *Journal of Food Science*, 80: C1170–C1177.

Engen, T. 1972 The effect of expectation on judgments of odor. *Acta Psychologica*, 36, 450–458.

Ewing, B. 2017. *Cider & Perry Organoleptic Profile Form*. Cider and Perry Workshops, Washington State University, Mount Vernon, WA.

Ewing, B.L., G.M. Peck, S. Ma, A.P. Neilson, and A.C. Stewart. 2019. Management of apple maturity and postharvest storage conditions to increase polyphenols in cider. *HortScience*, 54: 143–148.

Fan, W., Y. Xu, and A. Yu. 2006. Influence of oak chips geographical origin, toast level, dosage and aging time on volatile compounds of apple cider. *Journal of the Institute of Brewing*, 112: 255–263.

Forshey, C.G. 1976. *Training and Pruning Apple Trees*. Cornell Cooperative Extension Publishing. Information Bulletin 112, https://ecommons.cornell.edu/bitstream/handle/1813/17817/IB%20112.pdf?sequence=2&isAllowed=y, accessed May 5, 2019.

Galinato, S.P. and C.A. Miles. 2017. *Cost Estimates of Establishing and Producing Specialty Cider Apples in Central Washington*. Washington State Univ. Ext. Publ. TB35.

Galinato, S.P., R.K. Gallardo, and C.A. Miles. 2014. *Cost Estimation of Establishing a Cider Apple Orchard in Western Washington*. Washington State Univ. Ext. Publ. FS141E.

Girard, B. and O.L. Lau. 1995. Effect of maturity and storage on quality and volatile production of "Jonagold" apples. *Food Research International*, 28: 465–471.

Goldstein, E. and J. Goldstein. 2006. *Perfect Pairings: A master Sommelier's Practical Advice for Partnering Wine with Food*. Univ of California Press, Berkeley and Los Angeles, CA.

Gray, H. 1918. *Anatomy of the Human Body*. Lea and Febiger, Philadelphia, PA.

Green, B.G. 1996. Chemesthesis: pungency as a component of flavor. *Trends in Food Science & Technology*, 7: 415–420.

Green, B.G., J. Lim, F. Osterhoff, K. Blacher, and D. Nachtigal. 2010. Taste mixture interactions: suppression, additivity, and the predominance of sweetness. *Physiol Behav.*, Dec. 2, 101(5): 731–737.

Heikefelt, C. 2011. Chemical and sensory analyses of juice, cider and vinegar produced from different apple varieties. MS Thesis, Swedish Univ. of Agricultural Sciences, Alnarp.

Henderson, J.P. and D. Rex. 2012. *About Wine*, 2nd edn. Delmar Cengage Learning. Clifton Park, NY.

Hewson, L., T. Hollowood, S. Chandra, and J. Hort. 2009. Gustatory, olfactory and trigeminal interactions in a model carbonated beverage. *Chemosensory Perception*, 2: 94–107.

Institute of Food Technologists. 1975. *Minutes of Sensory Evaluation Div. Business Meeting*, at 35th Annual Meeting, Institute of Food Technologists, Chicago, IL.

Jackson, R.S. 2009. *Wine Tasting: A Professional Handbook*. Academic Press, Cambridge, MA.

Jolicoeur, C. 2013. *The New Cider Maker's Handbook. 2013*. Chelsea Green Publishing, White River Junction, VT.

Joslyn, M.A. and J.L.Goldstein. 1964. Astringency of fruits and fruit products in relation to phenolic content. *Adv. Food Res.*,13: 179–217.

Killian, E. and C.S. Ough. 1979. Fermentation esters – formation and retention as affected by fermentation temperature. *American Journal of Enology and Viticulture*, 30: 301–305.

Kim, S. and B. Lecat. 2017. An exploratory study to develop Korean food and wine pairing criteria. *Beverages*, 3: 40.

King, M. and M. Cliff. 2005 Evaluation of ideal wine and cheese pairs using a deviation-from-ideal scale with food and wine experts. *J. of Food Quality*, 28: 245–256.

Koyama, N. and K. Kurihara. 1972. Mechanism of bitter taste reception: interaction of bitter compounds with monolayers of lipids from bovine circumvallate papillae. *Biochimica et Biophysica Acta – Biomembranes* 1972, 288: 22–26.

Laaksonen, O., R. Kuldjärv, T. Paalme, M. Virkki, and B. Yang. 2017. Impact of apple cultivar, ripening stage, fermentation type and yeast strain on phenolic composition of apple ciders. *Food Chemistry*, 233: 29–37.

Lawless, H.T. and H. Heymann. 2010. *Sensory Evaluation of Food Principles and Practices*, 2nd edn. Springer Publishing Company, New York, NY.

Lea, A.G.H. 2014. Cidermaking, in *The Oxford Handbook of Food Fermentations* (C.W. Bamforth and R.E. Ward, eds), pp. 148–198. Oxford University Press. New York, NY.

Lea, A.G.H. and F.W. Beech. 1978. The phenolics of ciders: Effects of cultural conditions. *J. Sci. Food Agr.* 29:493–496.

Lee, C.B. and H.T. Lawless. 1991. Time-course of astringent sensations. *Chemical Senses*, 16: 225–238.

Lynch, P. 2006. *Medical Illustrations*. Center for Advanced Instructional Media, Yale University School of Medicine, New Haven, CT.

Malletroit, V., J.-X. Guinard, R.E. Kunkee, and M.J. Lewis. 1990. Effect of pasteurization on microbiological and sensory quality of white grape juice and wine. *Journal of Food Processing and Preservation*, 15:19–29.

Martineau, B., T. Henick-Kling, and T. Acree. 1995. Reassessment of the influence of malolactic fermentation on the concentration of diacetyl in wines. *American Journal of Enology and Viticulture*, 46: 385–388.

McMahon, K.M., C. Culver & C.F. Ross (2017) The production and consumer perception of sparkling wines of different carbonation levels, *Journal of Wine Research*, 28:2, 123–134.

Meilgaard, M.C., D.C. Reid, and K.A. Wyborski. 1982. Reference standards for beer flavor terminology system. *J. Am. Soc. Brew. Chem.*, 40: 119–128.

Merwin, I.A., S. Valois, and O. Padilla-Zakour. 2008. Cider apples and cider-making techniques in Europe and North America. *Hort. Rev.*, 34: 365–414.

Miles, C., J. King, T.R. Alexander and E. Scheenstra. 2017. Evaluation of flower, fruit, and juice characteristics of a multinational collection of cider apple varieties grown in the U.S. Pacific Northwest. *HortTechnology*, 27: 431–439.

Mitchell, P. 2016. *Cider & Perry Production - A Foundation. The Official textbook for the Foundation Certificate in Cider and Perry Production*. 1st Ed. The Cider & Perry Academy.

Molina, A.M., J.H. Swiegers, C. Varela, I.S. Pretorius, and E. Agosin. 2007. Influence of wine fermentation temperature on the synthesis of yeast-derived volatile aroma compounds. *Applied Microbiology and Biotechnology*, 77: 675–687.

Moncrieff, R.W. 1964. The metallic taste. *Perfumery Essent. Oil Rec.*, 55: 205–207.

Morrot, G. 2001. The color of odors. *Brain and Language*, 79, 309–320.

Moulton, G.A. and J. King. 2015. *Apple Varieties for Cooking, Baking & Cider*. Washington State University, https://extension.wsu.edu/maritimefruit/apple-varieties-for-cooking-baking-cider/, accessed August 13, 2019.

Moulton, G.A., C.A. Miles, J. King, and A. Zimmerman. 2010. *Hard Cider Production & Orchard Management in the Pacific Northwest*. Washington State Univ. Ext. Publ. PNW621.

NASS (National Agricultural Statistics Service). 2017. *Washington Tree Fruit Acreage Report 2017*. United States Department of Agriculture, https://www.nass.usda.gov/Statistics_by_State/Washington/Publications/Fruit/2017/FT2017.pdf, accessed May 5, 2019.

NWCA (Northwest Cider Association), Portland State University, and Irvine and Company LLC. 2016. *Northwest Cider Survey 2015*. Northwest Cider Association, Portland, OR.

Orton, V. 1995. *The American Cider Book: The Story of America's National Beverage*. North Point Press, New York, NY.

Parr, W.V., K.G. White and D.A. Heatherbell. 2003. The nose knows: influence of colour on perception of wine aroma. *Journal of Wine Research*, 14, 2–3: 79–101.

Peck, G., M. McGuire, T. Boudreau IV, and A. Stewart. 2016. Crop load density affect "York" apple juice and hard cider quality. *HortScience*, 51: 1098–1102.

Peng, B., T. Yue, and Y. Yuan. 2007. A fuzzy comprehensive evaluation for selecting yeast for cider making. *International Journal of Food Science and Technology*, 43: 140–144.

Petignat-Keller, S. 2013. *Flavour Wheel for Apple Juice and Cider*. Agroscope Changins-Wädenswil Research Station, Wädenswil, Switzerland.

Petrillo, N. 2016. *IBIS World Industry Report OD5335 Cider Production in the US*. IBISWorld. Los Angeles, CA.

Peynaud, E. 1980. *Le Goût du Vin*. Bordas, Paris.

Pinsley, K. 2019. Optimization of Hard cider processing to maximize tannin extraction: exploring the application of pulsed electric field to apple mash and pomace. Master's thesis, Cornell University.

Pisarnitskii, A.F. 2001. Formation of wine aroma: tones and imperfections caused by minor components (review). *Applied Biochemistry and Microbiology*, 37: 552–560.

Pontrelli, K. 2018. Cider-loving founding fathers. *CIDERCRAFT Magazine*, Seattle, WA https://cidercraftmag.com/2018/02/19/cider-loving-founding-fathers/, accessed October 16, 2017.

Salinger, S.V. 2002. *Taverns and Drinking in Early America*. Johns Hopkins University Press, Baltimore, MD.

Sánchez, A., de Revel, G., Antalick, G., Herrero, M., García, L.A., and Díaz, M. 2014. J Ind Microbiol Biotechnol 41: 853.

Simon, J. 1997. *Wine with Food*. Simon & Schuster, New York, NY.

Singleton, V.L., H.A., Sieberhagen, P. de Wet, and C.J. van Wyk. 1975. Composition and sensory qualities of wines from white grapes by fermentation with and without grape solids. *American Journal of Enology and Viticulture*, 26: 62–69.

Smith, D.V. and R.F. Margolskee. 2001. Making sense of taste. *Scientific American*, 284: 32–39.

Stangor, C. 2010. *Introduction to Psychology*. FlatWorld Knowledge LLC, Irvington, NY.

Statista, 2017. *Number of Cider Producers in the United States as of September 2016, by State*. Statista Inc., New York, NY, https://www.statista.com/statistics/300851/us-number-of-cider-manufacturers-by-state/, accessed August 23, 2017.

Stein, J.F and C.J. Stoodley. 2006. *Neuroscience: An Introduction*. John Wiley & Sons Inc, Hoboken, NJ.

Stephenson, M. 2018. *Cooking with Apple Cider Vinegar Cookbook*. Amazon Digital Services LLC, Kdp Print US.

Su, S.K. and R.C. Wiley. 1998. Changes in apple juice flavor compounds. *Journal of Food Science*, 63: 688–691.

Van Boekel, M.A.J.S. 2006. Formation of flavour compounds in the Maillard reaction. *Biotechnology Advances*, 24: 230–233.

Vidal, S., L. Francis, P. Williams, M. Kwiatkowski, R. Gawel, V. Cheynier, and E. Waters. 2004. The mouth-feel properties of polysaccharides and anthocyanins in a wine like medium. *Food Chemistry*, 85: 519–525.

Watson, B. 1999. *Cider Hard and Sweet: History, Traditions, and Making Your Own*. The Countryman Press, Woodstock, VT.

Wibowo, D., R. Eschenbruch, C.R. Davis, G.H. Fleet, and T.H. Lee. 1985. Occurrence and growth of lactic acid bacteria in wine: a review. *American Journal of Enology and Viticulture*, 36: 302–313.

Williams, A.A. 1974. Flavour research and the cider industry. *Journal for the Institute of Brewing*, 80: 455–470.

Williams, D.R. 1990. Hard cider's mysterious demise. George Mason Univ., Fairfax, VA, http://mason.gmu.cdu/~drwillia/cider.html, accessed February, 23, 2017.

Williams, R.R. 1975. *An Introduction to Modern Cider Apple Production*. Long Ashton Research Station, Bristol, England.

Williams, R.R. 1987. *Bulmer's Pomona*. Fourth Estate Ltd, London.

Zimmerman, A., G. Moulton, and C.A. Miles. 2015. *Fermentation Protocol at WSU Mount Vernon NWREC for Production of Varietal Ciders*. Washington State Univ., http://ext100.wsu.edu/maritimefruit/wp-content/uploads/sites/36/2015/04/CiderFermentationProtocol2015.pdf, accessed May 3, 2017.

Zoecklein, B.W., K.C. Fugelsang, B.H. Gump, F.S. Nury. 1990. *Production Wine Analysis*. Van Nostrand Reinhold, New York, NY.

Index

Note: bold page numbers indicate figures; italic page numbers indicate tables.

81

CABI – who we are and what we do

This book is published by **CABI**, an international not-for-profit organisation that improves people's lives worldwide by providing information and applying scientific expertise to solve problems in agriculture and the environment.

CABI is also a global publisher producing key scientific publications, including world renowned databases, as well as compendia, books, ebooks and full text electronic resources. We publish content in a wide range of subject areas including: agriculture and crop science / animal and veterinary sciences / ecology and conservation / environmental science / horticulture and plant sciences / human health, food science and nutrition / international development / leisure and tourism.

The profits from CABI's publishing activities enable us to work with farming communities around the world, supporting them as they battle with poor soil, invasive species and pests and diseases, to improve their livelihoods and help provide food for an ever growing population.

CABI is an international intergovernmental organisation, and we gratefully acknowledge the core financial support from our member countries (and lead agencies) including:

UKaid
from the British people

Ministry of Agriculture
People's Republic of China

Australian Government
Australian Centre for
International Agricultural Research

Agriculture and
Agri-Food Canada

Ministry of Foreign Affairs of the
Netherlands

Schweizerische Eidgenossenschaft
Confédération suisse
Confederazione Svizzera
Confederaziun svizra

Swiss Agency for Development
and Cooperation SDC

Discover more

To read more about CABI's work, please visit: **www.cabi.org**

Browse our books at: **www.cabi.org/bookshop**,
or explore our online products at: **www.cabi.org/publishing-products**

Interested in writing for CABI? Find our author guidelines here:
www.cabi.org/publishing-products/information-for-authors/